Cybersecurity: Cautionary Tales

Hacker Stories

Lessons Learned From the Dark Side of the Web

Dr. Sam Kurien

CHARIS CLARION PUBLISHING
www.charisclarion.com

Copyright

First published by **Charis Clarion Publishing House.**
Media/Charis Clarion Publishing House, Colorado Springs, CO, 80922
Printed in the USA

ISBN: 979-8-9946097-1-2
(Paperback Edition)
Library of Congress Cataloging-in-Publication Data
Kurien, Sam.
Cybersecurity: Cautionary Tales, HACKER STORIES - LESSONS
LEARNED FROM THE DARK SIDE OF THE WEB

Direction & Design: Simarjeet Kaur
Copyright & Layout: Isaac Kurien
Audio Narration: Thomas Jacobs

CHARIS CLARION PUBLISHING
www.charisclarion.com

Reviewed By:

Andrew Stravitz, CISSP, CISM, ITIL,
Founder and President of Touchpoint Cyber, LLC.
Virtual CISO services.

https://touchpointcyber.com

Former CISO in e-commerce, financial services, and critical infrastructure. Founder and president of Touchpoint Cyber, LLC, which provides IT Risk and Cybersecurity consultancy services to both small and large businesses.

This book could have been titled, "How not to be Hacked," as Sam provides not only a historically accurate account of the most impactful breaches and compromises of the past two decades, but also manages to place them in a useful taxonomy of 4 parts, which can serve as a go-to reference guide for any cybersecurity professional.

The case studies are carefully researched and laid out, which in itself could have been enough. However, Sam draws on decades of expertise to offer practical countermeasures and mitigation strategies that serve as a checklist to help security professionals and the general public understand the critical security threats of our current digital age.

Just as an astronaut has a checklist to ensure a positive outcome on each mission, the well-researched recommendations in the book will serve as critical templates for C-Suite executives, including CISOs and heads of audit. Sam doesn't just tell you that MFA is necessary; he also explains what to avoid (e.g., stop using SMS for 2FA) and recommends where companies need to migrate to strengthen their security posture.

Unlike many technical cybersecurity books, this material is presented in a conversational style, providing critical information for small businesses and individuals. The end of each chapter provides a clear, concise breakdown of what "security professionals" need to focus on, as well as what "everyone else" needs to consider and implement as part of their cyber hygiene.

This book should be required reading for aspiring cybersecurity professionals during their academic years, preparing them for the often chaotic realities they will soon face and providing proven strategies to mitigate malicious threats. I wish this could be included as supplemental reading in college Bachelor's or master's cybersecurity courses. I would have benefited from it.

Ultimately, every cybersecurity professional is focused on protecting the confidentiality, integrity, and availability of systems for both their employer and their customer base. This book draws on decades of prior knowledge and synthesizes it into a more straightforward path to success.

Felix Rajan Meri Cruze, CTO
Boscosoft Technologies

This book is an exceptionally well-crafted and timely contribution to the field of cybersecurity, distinguished by both its depth of technical insight and its universal relevance. What makes this work truly outstanding is the author's deliberate effort to ensure that every chapter delivers Lessons for Security Professionals as well as Lessons for Everyone. This dual-focus approach bridges the often-existing gap between technical expertise and public awareness, making the book both inclusive and impactful.

The chapters such as Ransomware Rampage, Vulnerabilities That Hurt Us All, and Digital Hygiene for Everyone are particularly eye-opening. They effectively expose how our everyday digital behaviors—whether performed knowingly or unknowingly—can create significant security risks. The author succeeds in transforming complex cybersecurity concepts into clear, relatable narratives that compel readers to reflect critically on their own digital habits.

The exploration of AI & Cybercrime is especially commendable. It highlights the evolving threat landscape with clarity and urgency, while also addressing the ethical responsibilities that come with technological advancement. Likewise, Our Identities at Risk powerfully underscores the fragility of digital identities in today's interconnected world and reinforces the need for vigilance at both individual and organizational levels.

One of the strongest aspects of this book is its consistent emphasis on preventive and ethical actions over corrective measures. The author convincingly advocates for proactive security practices, enforced policies, and disciplined digital behavior as the most effective means of protecting data and systems. This preventive mindset is precisely what is needed in an era where reactive security is no longer sufficient.

Overall, this book is not just informative—it is transformative. It challenges readers to rethink how they engage with technology and inspires a more responsible, ethical, and security-conscious digital culture. Whether you are a cybersecurity professional, a policymaker, or an everyday digital user, this book serves as a powerful reminder that protecting our digital world requires collective awareness, thoughtful policies, and consistent effort. It is, without question, a must-read and a significant contribution to modern cybersecurity literature.

Book Contents

Acknowlegdements

To my mom, whose prayers and love have brought me this far.

To the cybersecurity community at large—the defenders working tirelessly to protect our digital infrastructure—thank you for the work you do every day. Special thanks to leaders and friends who have served with me. Kris Rickert, as Network Operations Director, your support in securing infrastructure at the Alliance was pivotal. Loved the days and overnight work your team put in while responding to breaches. Jason Linscombe, for your watchful eyes on websites and web apps, and for strengthening perimeters, who provided perspective on the real-world impacts of these threats and reviewed sections for technical accuracy.

Foreword

Christopher Mosby
CEO, Movaci
CISSP, CCSP, CISA, CEH, CHFI, SecurityX, PCIP

https://movaci.com/

Cybersecurity has never been more complex or more personal than it is today. Every organization, regardless of size or sector, is navigating a world where the threats evolve faster than most teams can respond. At the same time, attackers are no longer limited to nation-state units or organized crime groups. Teenagers with curiosity and time can cause global damage. Criminal markets operate with the efficiency of modern startups. And every one of us is now a potential target simply because we exist in a connected world.

That is why this book matters. Cybersecurity: Cautionary Tales does not rely on hype or marketing language. It tells real stories that show how these incidents happen, why they happen, and what they cost. It also reminds us that cybersecurity is not just a technical discipline. It is a human one. Behind every breach, there are decisions, assumptions, blind spots, and incentives that make attacks possible long before any line of code is written.

I have spent much of my career working with organizations that care deeply about protecting their people and their mission. Along the way, I have worked with many leaders who understand the realities of security, but Sam stands out. I first met him when my team served as one of his cybersecurity vendors. I saw up close how he approached security from the inside out. It was never about buying a tool or checking a box. He understood that real security requires culture, training, governance, and accountability long before it requires technology. His experience across the commercial and non-profit sectors gives him a rare combination of technical depth, operational leadership, and practical judgment. This background allows him to speak clearly to the human and technical challenges most organizations face today.

This perspective is what makes the book valuable. These stories help the reader see patterns that repeat again and again. Human error. Misplaced trust. Overconfidence in technology. Underinvestment in people. A misunderstanding of how threat actors think. When you read the chapters on social engineering, SIM swaps, dark markets, or insider risk, you see how small decisions can escalate into major incidents. You also see why security teams cannot afford to operate in isolation from the rest of the organization. This foreword is not meant to summarize the book. The stories speak for themselves. Instead, my encouragement is simple. Read this with the mindset of a practitioner. Ask what lesson each chapter offers. Ask what assumptions it challenges. Ask where your own organization might share the same blind spots. If you do, you will come away with more than an understanding of the dark side of the internet. You will come away with a clearer sense of what it actually takes to build resilience.

The threats will continue to grow in speed and sophistication. That part is unavoidable. What we can control is how we respond, how we prepare our teams, and how we develop a culture that understands the human realities behind every system we defend. This book is a strong step in that direction.

Author's Note

Dr. Sam Kurien
CEO/CIO and VP of Technology
CIO at Denver Seminary, Formerly the Global CIO at The Alliance and CEO of consulting and development firms,

bitsbytesandbricks.blogspot.com

We live in two different digital worlds. In one, security professionals battle invisible enemies using acronyms and attack vectors. In the other, everyday people click links, share passwords, and trust systems they don't understand. These worlds rarely communicate effectively, yet both face the same threats.

I wrote this book to bridge that dangerous gap.

As CEO/CIO and VP of Technology with various organizations over the last two decades

What These Stories Offer

Security professionals will discover the human psychology driving attacks, learn from real incident responses, and develop better ways to communicate complex threats to non-technical audiences. Each chapter concludes with specific technical lessons you can implement immediately.

Everyone else will learn to recognize attack warning signs, understand what to do when compromised, and protect their families online. Every chapter includes a "Lessons for Everyone" section, translating technical concepts into practical advice you can use today.

Why Stories Matter

I chose narrative over technical manual because threats aren't abstract—they have faces, motivations, and consequences. The teenager arrested for SIM swapping was someone's child. The hospital administrator paying ransom was trying to save lives. The security researcher with a criminal past was seeking redemption.

These twelve stories—from Kevin Mitnick's social engineering to modern hospital ransomware—aren't just history. They're patterns that repeat with each new technology. Understanding them helps us predict and prevent future attacks. While some names have been changed for privacy, these accounts are based on real incidents documented in court records and security reports. Any errors in interpretation are mine alone.

We're all connected in this digital age. A vulnerability in one system can cascade globally within hours. But so can knowledge, awareness, and resilience. These aren't just cautionary tales—they're survival guides for the world we now inhabit.

Whether you're defending a Fortune 500 company or just trying to protect your family's photos, these stories have lessons for you. Stay vigilant, stay informed, but most importantly, stay human. Our greatest vulnerability and our greatest strength remain the same: each other.

Chapter Contents Part-1

PART I - HACKERS AND SOCIAL ENGINEERING

The Human Factor - Kevin Mitnick's Social Engineering Legacy

- The Most Wanted Hacker in America
- The Art of Deception
- Pacific Bell and COSMOS
- The FBI Pursuit
- From Black Hat to White Hat
- Lessons for Security Professionals
- Lessons for Everyone

The Teenage Bank Robber - Digital Age Financial Attacks

- The Bedroom Hacker
- Evolution from Curiosity to Crime
- The Commonwealth Bank Breach
- International Investigation
- The Price of Digital Crime
- Lessons for Security Professionals
- Lessons for Everyone

The SIM Swap Scammers - When Your Phone Becomes Your Enemy

- Michael Terpin's $24 Million Loss
- The OG Users
- Carrier Complicity
- Law Enforcement Response
- The Ongoing Threat
- Lessons for Security Professionals
- Lessons for Everyone

PART II – MARKETS, LEAKS, AND MOVEMENTS

Dark Markets - The Rise and Fall of Digital Black Markets

- The Capture of Ross Ulbricht
- Building the Silk Road
- The Libertarian Dream Corrupted
- Law Enforcement Strikes Back
- The Hydra Effect
- Lessons for Security Professionals
- Lessons for Everyone

Inside Anonymous - The Hacktivist Phenomenon

- Operation Payback
- From 4chan to Global Movement
- Project Chanology and the Guy Fawkes Mask
- Arab Spring and Global Protests
- LulzSec and the Fall
- Lessons for Security Professionals
- Lessons for Everyone

The Insider Threat - When Trust Becomes Vulnerability

- Edward Snowden's Decision
- The Path to Betrayal
- The Scope of Surveillance
- Global Revelations
- The Price of Truth
- Lessons for Security Professionals
- Lessons for Everyone

Chapter Contents Part-3

PART IV – VULNERABILITIES THAT HURT US ALL

The Hospital Hack - When Ransomware Threatens Lives

- Death in Düsseldorf
- Healthcare's Perfect Storm
- Universal Health Services Under Attack
- The Human Cost
- The Underground Economy
- Lessons for Security Professionals
- Lessons for Everyone

Greg - The Youngest Hacker's Path from Prodigy to Prison

- The FBI Raid
- From Gamer to Criminal
- The Discord Crew
- Arrest and Consequences
- Lost Potential
- Lessons for Security Professionals
- Lessons for Everyone

MalwareTech - The Accidental Hero Who Stopped WannaCry

- Stopping a Global Attack
- The Double Life
- Kronos and Redemption
- From Hero to Defendant
- The Complexity of Second Chances
- Lessons for Security Professionals
- Lessons for Everyone

Part-I

Hackers and Social Engineering

Introduction to Part I

In the digital age, the most sophisticated firewalls, the most complex encryption algorithms, and the most expensive security systems all share a common vulnerability: the human beings who use them. Part I of this book explores how hackers have exploited this fundamental weakness, turning our trust, helpfulness, and technological dependence into weapons against us.

These stories, drawn from real incidents that have shaped the cybersecurity landscape, reveal an uncomfortable truth. While we've built increasingly complex technical defenses, the most devastating attacks often succeed not by breaking through our digital walls, but by convincing us to open the door from the inside. From Kevin Mitnick's pioneering social engineering techniques to modern SIM swap attacks that can drain bank accounts in minutes, these chapters trace the evolution of human-targeted hacking from curious exploration to organized crime.

Note: These topics have been extensively covered in cybersecurity media, including podcasts like Darknet Diaries by Jack Rhysider, security conferences, and investigative journalism. This narrative draws from publicly available sources to present these critical security stories.

Chapter-1

The Human Factor - Kevin Mitnick's Social Engineering Legacy

"The weakest link in the security chain is the human element. Technology is rarely the problem—people are."
— **Kevin Mitnick**, *after his transformation to security consultant*

February 15, 1995, Raleigh, North Carolina

The knock came at 1:30 AM, sharp and authoritative against the apartment door. Kevin Mitnick, America's most wanted computer hacker, knew instantly that his two-and-a-half-year run was over. As FBI agents flooded into his sparse apartment, finding little more than a laptop, some clothes, and a mattress on the floor, they were arresting not just a man but a legend—someone who had penetrated some of the most secure computer systems in the world armed primarily with a telephone and an uncanny understanding of human psychology.

The apartment's spartanness belied the sophistication of what Mitnick had accomplished. No walls covered in monitors displaying scrolling green code, no stacks of hardware, no evidence of the Hollywood hacker stereotype. Just a tired-looking thirty-one-year-old man who had proven that the most powerful hacking tool wasn't a computer program—it was the human voice.

Mitnick's journey to that moment had begun decades earlier in Los Angeles, where a lonely, overweight kid discovered that computers—unlike people—behaved predictably. They followed rules. They made sense. But his real education came not from computers but from phone phreakers, the telephone system hackers who taught him that every system, no matter how complex, was designed, maintained, and operated by human beings. And human beings, Mitnick learned, were far easier to hack than machines.

His methodology was devastatingly simple and remarkably sophisticated. Where other hackers spent weeks trying to crack passwords or exploit technical vulnerabilities, Mitnick would simply call a company's help desk. "Hi, this is Tom from the Denver office," he'd say, his voice carrying just the right mixture of authority and frustration. "I'm locked out of my account and I've got a presentation to the board in an hour. Can you reset my password?" More often than not, they would.

The technical term for what Mitnick pioneered is "social engineering," but that clinical phrase fails to capture the artistry involved. Each call was a performance, carefully scripted yet naturally delivered. Mitnick studied organizational charts, learned corporate jargon, and understood the psychology of help desk workers—overworked, undertrained, and measured by how quickly they resolved issues. He knew that creating a sense of urgency short-circuited critical thinkg. He understood that people naturally wanted to help, especially when the caller seemed to be a fellow employee in distress.

His infiltration of Pacific Bell's COSMOS system demonstrated the power of this approach. COSMOS controlled phone service for California, containing unlisted numbers, the ability to create new lines, and access to call records. Traditional hackers might have spent months probing for technical vulnerabilities. Mitnick spent a few days doing reconnaissance, learning the names of employees, understanding the system's purpose, and most importantly, identifying the engineers who had access.

Then he made his calls. First to human resources, posing as a new employee who hadn't received his orientation packet. Could they send him the internal phone directory? Next to IT support, claiming to be that same new employee having trouble accessing the network. What was the format for usernames again? Within a week, he had enough information to convincingly impersonate a Pacific Bell engineer. When he finally called the COSMOS administrators, he was no longer Kevin Mitnick—he was Steve from the San Francisco office, dealing with a critical service issue, and he needed access immediately.

The administrators never questioned him. Why would they? He knew the right names, used the correct terminology, and most importantly, he sounded like he belonged. Within minutes, he had login credentials to one of the most sensitive systems in California's telecommunications infrastructure. No code was written, no vulnerabilities exploited—just human trust weaponized.

But Mitnick's most audacious hack targeted Digital Equipment Corporation (DEC), one of the world's largest computer companies. His goal was their VMS operating system source code—the crown jewels of a multi-billion dollar corporation. Again, technical attacks weren't necessary. Through a series of phone calls, he convinced DEC employees that he was a fellow worker collaborating on a project. He learned about their development systems, their security procedures, and eventually, how to dial directly into their internal network.

The source code he downloaded was worth millions, representing years of development by hundreds of programmers. Yet Mitnick never sold it, never profited from it. For him, the hack itself was the reward—proof that he could penetrate any system, not through superior technology but through superior understanding of human nature.

His techniques evolved with technology. As companies implemented callback systems—where the system would hang up and call back a pre-authorized number—Mitnick learned to manipulate phone switches to redirect the callbacks. When organizations required verbal passwords, he would social engineer the passwords themselves, often from the very people who created them. Every security measure designed to stop technical attacks became another human vulnerability to exploit.

The FBI's pursuit of Mitnick became one of the most extensive computer crime investigations in history. Agent Ken McGuire later described the frustrationof chasing someone who seemed to be everywhere and nowhere simultaneously. Mitnick monitored the agents hunting him, accessing their emails and voicemails, staying perpetually one step ahead. He used cloned cell phones to avoid detection, switching identities as easily as changing clothes.

The investigation revealed the scope of Mitnick's infiltrations: Motorola, Nokia, Sun Microsystems, Fujitsu, Novell—a who's who of technology companies had been compromised. Yet in each case, the initial breach came not through sophisticated malware or zero-day exploits, but through employees who believed they were helping a colleague, a vendor, or a customer.

Tsutomu Shimomura, the computational physicist whose pursuit ultimately led to Mitnick's capture, noted something remarkable: despite having access to systems that controlled millions of dollars, Mitnick never enriched himself financially. He could have sold corporate secrets, stolen money, or destroyed critical infrastructure. Instead, he collected access like trophies, each successful infiltration proving his mastery of the most complex security system of all—human trust.

The technology community's reaction to Mitnick's arrest was polarized. Many saw him as a dangerous criminal who had violated the trust that made modern business possible. Others viewed him as a pioneer who had exposed fundamental vulnerabilities that needed addressing. The hacker community, in particular, was divided between those who admired his skills and those who believed he had crossed ethical lines that damaged the reputation of all security researchers.

During his trial, the prosecution struggled to quantify damages. How do you value source code that was copied but never distributed? How do you calculate the cost of access that was gained but never maliciously used? The government claimed millions in damages, while Mitnick's supporters argued that the only real harm was embarrassment to corporations that had failed to implement basic security procedures.

The judge's sentence—five years in federal prison, including eight months in solitary confinement—reflected the era's fear of computer hackers. Mitnick was denied access to telephones for years, based on the prosecution's claim that he could "start a nuclear war by whistling into a pay phone"—a technically impossible but psychologically revealing assertion about how society viewed hackers.

Upon his release in 2000, Mitnick faced three years of supervised release during which he was prohibited from touching any computer or cell phone. For someone whose entire identity was intertwined with technology, it was a form of digital death. Yet this period of forced reflection transformed him. When his restrictions were finally lifted in 2003, Mitnick emerged not as a black hat hacker but as one of the world's most sought-after security consultants.

His transformation from hacker to security consultant wasn't just a career change—it was a fundamental shift in how the industry approached security. Mitnick's company began offering "social engineering penetration testing," where his team would attempt to breach organizations using the same techniques he had pioneered. The results were consistently devastating. Despite years of warnings and training, employees still fell for the same tactics Mitnick had used in the 1990s.

In one memorable test, Mitnick's team was hired to evaluate a banks security. Within hours, they had convinced an employee to install a "security update" that was actually malware. Within days, they had complete access to the bank's internal network.

The bank had spent millions on technical security but had been compromised by a single phone call and a convincing email.

These demonstrations forced a fundamental reevaluation of security priorities. Organizations began to realize that their security budgets, overwhelmingly focused on technical controls, had ignored their greatest vulnerability. Security awareness training, once an afterthought, became mandatory. Companies implemented verification procedures for sensitive requests. The concept of "zero trust"—assume every request is potentially malicious until proven otherwise—began to take hold.

Yet even as organizations adapted, social engineering evolved. The same techniques Mitnick pioneered were adopted by criminals worldwide, refined and weaponized for financial gain. The "CEO fraud" scams that now cost businesses billions annually are direct descendants of Mitnick's methods. The phone calls from "Microsoft support" that plague consumers globally follow his playbook. The spear-phishing emails that compromise major corporations use the same psychological principles he demonstrated.

Modern social engineers have advantages Mitnick never enjoyed. Social media provides a wealth of information about targets—their interests, relationships, and daily activities. LinkedIn reveals organizational structures and employee roles. Facebook exposes personal details that can make impersonation convincing. The digital breadcrumbs we all leave online have made reconnaissance trivially easy.

Artificial intelligence has added new dimensions to social engineering. Deepfake audio can perfectly mimic voices, making phone-based attacks even more convincing. Natural language processing helps attackers craft persuasive emails at scale. Machine learning algorithms can analyze thousands of potential targets to identify the most vulnerable.

Yet for all these technological advances, the fundamental vulnerability remains unchanged: human beings want to be helpful. We're programmed by evolution and society to cooperate, to trust, to assist those in need. These traits, essential for civilization, become weaknesses in the face of malicious manipulation.

Mitnick, now in his sixties, continues to demonstrate these vulnerabilities at security conferences worldwide. His presentations are part education, part performance art, as he shows audiences how easily they can be manipulated. He'll call a volunteer's bank live on stage, extracting information through nothing more than confidence and preparation. The audience watches in mixed fascination and horror as security measures crumble before their eyes.

"Technology has changed," Mitnick often says, "but people haven't." He points out that every major breach in recent years—from the Target hack to the SolarWinds incident—involved social engineering at some stage. Attackers may use sophisticated malware, but they deliver it through human manipulation. They may exploit zero-day vulnerabilities, but they discover them through employees who inadvertentl reveal too much.

The legacy of Kevin Mitnick extends far beyond his individual exploits. He fundamentally changed how we think about security, forcing recognition that the human element isn't just another attack vector—it's often the primary one. His story serves as both a cautionary tale and a roadmap, showing how human nature can be exploited while also pointing toward defenses.

Today's security professionals study Mitnick's techniques not as historical curiosities but as active threats. Every security assessment now includes social engineering tests. Every security training emphasizes the human factor. Every incident response plan accounts for the possibility that the breach began not with a technical failure but with a successful lie.

The transformation of Kevin Mitnick from hacker to security consultant mirrors the evolution of cybersecurity itself. What began as a game played by curious individuals has become a critical component of national security and economic stability. The kid who hacked for the thrill of it became the adult warning others about those who hack for profit or power.

As we face a future of increasingly sophisticated cyber threats, Mitnick's story reminds us of a fundamental truth: the most advanced firewall is useless if someone can convince an employee to hold the door open. The strongest encryption is meaningless if someone can persuade a user to share their password. The most secure system is vulnerable if someone can manipulate the people who operate it.

Lessons for Security Professionals

1. Implement Robust Verification Procedures
Establish multi-factor verification for all sensitive requests, especially password resets and access grants
- Create callback procedures using pre-verified contact information, not caller-provided numbers
- Implement challenge questions that only legitimate employees would know
- Require manager approval for unusual requests, regardless of claimed urgency
- Document all verification steps to create an audit trail

2. Develop Comprehensive Security Awareness Programs
- Conduct regular training that includes real-world scenarios and live demonstrations
- Test employees with simulated social engineering attacks and provide immediate feedback
- Create a security-conscious culture where questioning unusual requests is encouraged
- Share stories of actual breaches to make threats tangible and memorable
- Reward employees who successfully identify and report social engineering attempts

3. Technical Controls Against Social Engineering
- Deploy privileged access management (PAM) systems to limit nd monitor administrative access
- Implement segregation of duties so no single person can complete sensitive operations
- Use out-of-band verification for high-risk transactions
- Deploy user behavior analytics to detect unusual access patterns.
- Maintain detailed logs of all administrative actions for forensic analysis

4. Incident Response Preparedness
- Include social engineering scenarios in incident response plans
- Train response teams to investigate the human element of breaches
- Develop procedures for quickly disabling compromised accounts
- Create communication protocols that don't rely on potentially compromised channels
- Conduct post-incident reviews focusing on how social engineering was used

5. Zero Trust Architecture Implementation
- Assume all requests are potentially malicious until verified
- Implement least privilege access controls universally
- Require re-authentication for sensitive operations
- Monitor and log all privileged activities
- Regularly review and revoke unnecessary access permissions

Lessons for Everyone

1. Healthy Skepticism is Your Best Defense
- Question unexpected calls or emails, especially those creating urgency
- Verify requests through independent channels—hang up and call back using known numbers
- Remember that legitimate organizations won't threaten immediate consequences
- Be suspicious of anyone asking for passwords or sensitive information
- Trust your instincts—if something feels wrong, it probably is

2. Protect Your Personal Information
- Limit what you share on social media—every detail can be used against you
- Be cautious about personality quizzes and surveys that harvest personal data
- Use different passwords for different accounts to limit breach impact
- Enable two-factor authentication wherever possible
- Regularly review privacy settings on all online accounts

3. Recognize Common Manipulation Tactics
- Urgency: "This must be done immediately or terrible consequences will occur"
- Authority: "I'm from IT/the government/your bank and need your cooperation"
- Familiarity: "Hi, it's me, your friend/colleague/relative, I need help"
- Intimidation: "You'll be in serious trouble if you don't comply"
- Sympathy: "I'm in a difficult situation and only you can help"

4. Safe Communication Practices
- Never give passwords over phone or email—legitimate companies won't ask
- Verify unexpected money requests, even from known contacts
- Don't click links in emails—navigate to websites directly
- Be cautious about downloading attachments, even from known senders
- Use video calls to verify identity for sensitive requests

5. What to Do If You Suspect You've Been Targeted

- Don't panic—quick action can minimize damage
- Change passwords immediately for affected accounts
- Contact your bank if financial information was involved
- Report the incident to appropriate authorities (IC3.gov for internet crimes)
- Warn others who might be targeted using your information
- Document everything for potential investigation

Chapter-2

The Teenage Bank Robber - Digital Age Financial Attacks

"We're not dealing with masterminds here. We're dealing with kids who have too much time, too much skill, and too little supervision."
— *FBI Special Agent* **Michael Christman,** *on teenage financial hackers*

August 2018, Brisbane, Australia

The knock on Matthew's bedroom door came at 6 AM. The seventeen-year-old was still in his Spider-Man pajamas when Australian Federal Police officers entered his room, their eyes widening at the sophisticated setup before them. Three monitors displayed cascading terminal windows, cryptocurrency exchange interfaces, and what appeared to be active penetration testing tools. Post-it notes with IP addresses and passwords covered every available surface. A half-eaten pizza sat next to a collection of energy drinks and a well-worn copy of "The Art of Exploitation."

"Is this about the banks?" Matthew asked, his voice cracking with a mixture of adolescence and resignation. He'd been expecting this visit for weeks, watching the news coverage of the attacks he'd participated in, seeing the escalating rewards for information, knowing that somewhere in his trail of proxies and encrypted communications, he'd made a mistake.

The officers exchanged glances. This kid—and he was definitely still a kid, with acne and a nervous habit of pushing up his glasses—had been part of a group that had stolen over four million dollars from Australian financial institutions. He wasn't what they'd expected. There was no bravado, no attempt to destroy evidence, just a tired teenager who seemed almost relieved that it was over.

Matthew's story, like that of hundreds of teenage hackers worldwide, began not with criminal intent but with curiosity. At thirteen, he'd discovered a vulnerability in his school's network that allowed him to change grades. He didn't use it—the discovery itself was the thrill. By fourteen, he was participating in bug bounty programs, legally finding vulnerabilities in corporate systems for rewards. Google paid him $3,000 for discovering a flaw in their OAuth implementation. Microsoft sent him a letter of appreciation for responsibly disclosing a Windows vulnerability.

But somewhere between legitimate security research and criminal hacking, Matthew crossed a line he couldn't clearly define. It started in the forums—not the public ones where security professionals gathered, but the hidden ones accessible only through Tor, where usernames replaced real names and cryptocurrency replaced traditional payment methods. Here, teenagers from around the world gathered to share techniques, trade exploits, and eventually, plan attacks.

The forums were intoxicating for a lonely teenager. Matthew, who struggled with social interactions at school, found himself respected and sought after online. His technical skills, meaningless to his classmates obsessed with sports and parties, made him valuable in this digital underground. Users with handles like "PhantomSec" and "NullRoute" became his closest friends, even though he never knew their real names or faces.

The evolution from curiosity to crime was gradual. First, it was just sharing proof-of-concepts—demonstrating that vulnerabilities existed without exploiting them. Then came the "educational" tools—programs that could theoretically be used for attacks but were presented as research instruments. Finally, there were the propositions: "Want to make some real money?" "I have access to a bank's test environment, want to see if it works on production?" "No one gets hurt, it's all insured."

The technical sophistication of these teenage attackers astounded investigators. They weren't using simple phishing emails or basic malware. Matthew's group had developed custom exploit frameworks that rivaled nation-state tools. They understood complex financial systems better than many banking employees. They could navigate SWIFT networks, manipulate ACH transfers, and exploit weaknesses in banking APIs that even the banks didn't know existed.

Their attack on Commonwealth Bank began with reconnaissance that would make military intelligence proud. For months, they mapped the bank's digital infrastructure, identifying employees through LinkedIn, tracking network ranges through WHOIS lookups, and discovering forgotten development servers through search engine dorking. They monitored job postings to understand what technologies the bank used. They analyzed the bank's mobile app, reverse-engineering its API calls to understand backend systems.

The initial breach came through a vulnerability in a third-party vendor's payment processing system. The vendor, a small company that provided services to multiple banks, had far weaker security than the banks themselves. Matthew wrote the exploit—a beautiful piece of code that chained together three separate vulnerabilities to achieve remote code execution. He was proud of it, the way an artist might be proud of a painting.

Once inside the vendor's network, they moved laterally, using stolen credentials and exploiting trust relationships between systems. They installed persistent backdoors, carefully designed to mimic legitimate traffic. They exfiltrated data slowly, staying below detection thresholds. When they finally reached the bank's core systems, they had multiple entry points and complete situational awareness.

The actual theft was almost anticlimactic. Using their access, they created legitimate-looking transactions that moved money from dormant accounts to mule accounts they controlled. They timed the transfers for Friday afternoons, knowing that weekend processing delays would give them time to move the money through multiple jurisdictions. They used cryptocurrency tumblers to obscure the trail, converting stolen funds to Bitcoin, then Monero, then back to Bitcoin through dozens of exchanges.

Matthew's share was $400,000—more money than his parents had made in the last five years combined. But he couldn't spend it, couldn't even tell anyone about it. The cryptocurrency sat in wallets whose keys he had memorized, a fortune he was too terrified to touch. He watched his parents stress about mortgage payments while he possessed enough stolen money to pay off their house ten times over.

The psychological toll was devastating. Matthew developed insomnia, lying awake imagining police raids. He became paranoid, convinced that every van on his street was surveillance, that every new student at school was an undercover officer. His grades plummeted. His parents, thinking he was depressed, sent him to therapy, where he sat in silence, unable to confess to the therapist who assured him that nothing he said would leave the room—except, she always added, if he was a danger to himself or others. Did stealing millions count as danger to others?

Meanwhile, the attacks continued. Other members of the group, emboldened by success, targeted larger institutions. They hit investment firms, payment processors, and cryptocurrency exchanges. The techniques evolved—using machine learning to identify dormant accounts, automating money muling through compromised business accounts, even attempting to manipulate market prices through strategic information leaks.

The banking industry's response was fragmented and inadequate. Each institution implemented security measures in isolation, unaware that attackers were sharing intelligence about their defenses. Banks that detected intrusions often kept quiet, fearing regulatory penalties and reputational damage. By the time they shared indicators of compromise, the attackers had evolved their techniques.

The age of the attackers complicated everything. When investigators finally identified suspects, they found children—legally minors in most jurisdictions. How do you prosecute a fifteen-year-old for crimes that would earn an adult decades in prison? How do you explain to shareholders that their bank was compromised by someone too young to have a driver's license?

Matthew's group made their fatal mistake through hubris. One member, a sixteen-year-old from the United States, couldn't resist bragging. He posted screenshots of his cryptocurrency wallets on Instagram, bought a Tesla with stolen funds, and wore a $50,000 watch to high school. When arrested, he quickly cooperated, providing usernames and technical details about their operations.

The international investigation that followed revealed the true scope of teenage financial hacking. Law enforcement agencies identified hundreds of teenagers involved in financial crimes across dozens of countries. Some worked in organized groups like Matthew's. Others operated alone, using tools and techniques shared in forums. The total losses ran into hundreds of millions, possibly billions—exact figures were impossible to determine given the number of undetected or unreported breaches.

The legal outcomes varied wildly by jurisdiction. In some countries, the teenagers were tried as adults and faced serious prison time. In others, they received suspended sentences and community service. Matthew, caught in themiddle ground, received two years in a juvenile facility and five years of probation during which he was banned from using computers—a sentence that effectively excluded him from the modern economy.

The reformation of teenage hackers became a contentious issue. Some, like Matthew, genuinely wanted to use their skills legitimately. They understood systems in ways that few security professionals did. Their experience attacking banks made them invaluable for defense. But who would hire someone with a criminal record for financial crimes to protect financial systems?

Some countries experimented with alternative approaches. The Netherlands created rehabilitation programs that channeled hacking skills toward legitimate careers. The UK developed cyber-rehabilitation initiatives that paired young offenders with mentors from the security industry. These programs showed promise, with some former teenage hackers becoming successful security researchers.

But for every success story, there were failures. Some teenage hackers, introduced to criminal networks through their activities, descended deeper into organized crime. Others, banned from using computers during critical years of technological development, found themselves unemployable in any modern field. The very skills that made them dangerous also made them valuable, creating a paradox that society struggled to resolve.

The financial industry, forced to confront its vulnerabilities, invested billions in security improvements. But the fundamental challenge remained: how do you protect against attackers who understand your systems better than you do? How do you defend against teenagers who have unlimited time, minimal risk perception, and treat your security as a game to be won?

Modern financial attacks by teenagers have evolved beyond simple theft. They manipulate cryptocurrency markets through coordinated pump-and-dump schemes. They exploit decentralized finance (DeFi) protocols, draining millions from smart contracts. They conduct ransomware attacks on financial infrastructure, threatening economic stability. The teenager in pajamas has become one of the financial industry's most serious threats.

The story of Matthew and his contemporaries forces us to confront uncomfortable questions about technology, youth, and crime. These aren't hardened criminals but children who found themselves with the power to commit massive crimes before they had the judgment to understand consequences. They're products of a society that celebrates hacking in movies and video games while failing to provide legitimate outlets for their skills.

Today, Matthew works as a barista, his computer ban finally lifted but his criminal record makes technology jobs impossible. He watches developments in cybersecurity with the intensity of an exiled citizen following news from home. Sometimes customers pay with phones that he knows he could compromise. Banking apps display on screens inches from his face, their vulnerabilities obvious to his trained eye. But he makes their coffee, takes their money, and says nothing.

Late at night, he sometimes logs into the old forums through Tor, not to participate but to watch. He sees new usernames, recognizes the same patterns, and watches teenagers make the same journey he did from curiosity to crime. He wants to warn them, to explain that the money isn't worth the cost, that the thrill fades but the consequences remain. But he knows they wouldn't

---

listen. He wouldn't have listened.

The financial industry continues to invest significant resources in defending against these young attackers. They hire consultants, implement new technologies, and conduct training. But somewhere, in a bedroom decorated with superhero posters, a teenager is discovering their first vulnerability, feeling that intoxicating rush of unauthorized access, taking their first step on a path that leads either to a legitimate career in security or to a morning visit from law enforcement.

The question isn't whether there will be more teenage bank robbers—there will be. The question is whether society can create paths that channel their skills productively before they cross lines that can't be uncrossed. The digital age has given teenagers unprecedented power. How we help them use it responsibly will determine whether they become the defenders or destroyers of our financial future.

Lessons for Security Professionals

1. Third-Party Risk Managemen
- Audit all vendor security practices regularly, not just during onboarding
- Implement continuous monitoring of third-party access and activities
- Require vendors to maintain security standards equal to your own
- Create vendor risk tiers based on access levels and data sensitivity
- Develop incident response plans that account for vendor compromises

2. Advanced Threat Detection
- Deploy behavioral analytics to identify unusual transaction patterns
- Implement deception technology (honeypots) to detect lateral movement
- Use machine learning to identify dormant account exploitation
- Monitor for data exfiltration patterns, not just large transfers
- Correlate security events across multiple systems and timeframes

3. Collaborative Defense Strategies
- Share threat intelligence with industry peers in real-time
- Participate in financial services information sharing initiatives (FS-ISAC)
- Coordinate responses to attacks targeting multiple institutions
- Develop industry-wide standards for API security
- Create joint task forces for investigating sophisticated attacks

4. Insider Threat Programs
- Monitor for unauthorized access to production systems from development environments
- Implement separation between test and production environments
- Track unusual patterns in privileged account usage
- Develop psychological profiles of potential insider threats
- Create anonymous reporting mechanisms for suspicious activities

5. Cryptocurrency and Digital Asset Security
- Understand blockchain analysis techniques for tracing stolen funds
- Develop relationships with cryptocurrency exchanges for freeze requests

- Implement controls for detecting cryptocurrency mining malware
- Monitor for unauthorized cryptocurrency transactions
- Create procedures for responding to cryptocurrency-related crimes

Lessons for Everyone

1. Understanding Your Digital Financial Footprint
- Regularly review all bank and credit card statements for unusual activity
- Set up alerts for all financial transactions, no matter how small
- Understand what normal looks like so you can spot abnormal
- Know which accounts are linked and how compromise of one affects others
- Keep records of all financial accounts, including dormant ones

2. Protecting Your Financial Identity
- Use unique, complex passwords for every financial account
- Enable two-factor authentication using apps, not SMS when possible
- Never access financial accounts on public WiFi
- Regularly check your credit reports for unauthorized accounts
- Consider freezing credit if you're not actively using it

3. Recognizing Financial Scams
- Banks will never ask for your full password or PIN
- Be suspicious of urgent requests to move money
- Verify any unexpected changes to your account through official channels
- Don't trust caller ID—it can be spoofed
- Question any request to install software for "security purposes"

4. Safe Online Banking Practices
- Always type your bank's URL directly, never click email links
- Look for HTTPS and verify the certificate is correct
- Log out completely when finished, don't just close the browser
- Use a dedicated device or browser for financial activities if possible
- Keep your banking app updated to the latest version

5. Responding to Financial Fraud

- Contact your bank immediately if you suspect any unauthorized activity
- Document everything—times, dates, amounts, and who you spoke with
- File a police report for any significant fraud
- Contact credit bureaus to place fraud alerts
- Consider identity theft protection services if you've been compromised
- Change passwords for all financial accounts, not just compromised ones

Chapter-3

The SIM Swap Scammers - When Your Phone Becomes Your Enemy

"Your phone number has become your identity, and it was never designed for that. We built our entire security infrastructure on a foundation of sand."
— **Michael Terpin,** *after losing $24 million to SIM swappers*

January 2019, Los Angeles, California

Michael Terpin stared at his phone in disbelief. The "No Service" message had appeared twenty minutes ago, but he'd initially dismissed it as a typical AT&T glitch. Now, as he watched his email inbox fill with password reset confirmations for accounts he wasn't accessing, the horrible reality set in. His phone number—the digital key to his entire online life—had been stolen.

Within minutes, twenty-four million dollars in cryptocurrency vanished from his accounts.

The attack on Terpin wasn't random. The perpetrators, a loose collective of teenagers calling themselves "OG Users," had spent weeks preparing. They knew he was a cryptocurrency investor. They knew which exchanges he used. They knew his phone number, his email addresses, and enough personal information to answer security questions. What they needed was control of his phone number, and they knew exactly how to get it.

The SIM swap attack represents a perfect storm of technical vulnerability, social engineering, and systemic weakness in our authentication systems. It exploits a fundamental flaw in how we've built digital identity: the assumption that control of a phone number equals identity verification. This assumption, hardcoded into countless services and systems, has created a single point of failure worth billions.

The technique itself is elegantly simple. Attackers gather information about their target through data breaches, social media, and public records—mothers' maiden names from genealogy sites, high school mascots from Facebook, last four digits of SSNs from previous breaches. Armed with this information, they contact the victim's mobile carrier, claiming to be the victim with a new phone. Through social engineering, bribery, or insider assistance, they convince the carrier to transfer the victim's number to a SIM card they control.

The moment the swap completes, the attacker receives all calls and texts meant for the victim. Two-factor authentication codes, password reset links, verification calls—everything flows to the attacker's device. The victim's phone, suddenly disconnected, becomes a useless piece of plastic and silicon. By the time most victims realize something is wrong, their digital lives have been systematically pillaged.

The rise of cryptocurrency transformed SIM swapping from a niche attack to a lucrative criminal enterprise. Traditional banking systems have fraud protections, transaction delays, and reversal mechanisms. Cryptocurrency transactions are immediate, irreversible, and pseudonymous. A successful SIM swap against a cryptocurrency holder could net millions in minutes with almost no possibility of recovery.

pseudonymous. A successful SIM swap against a cryptocurrency holder could net millions in minutes with almost no possibility of recovery.

The "OG Users" represented a new breed of criminal—teenagers who viewed SIM swapping as a competitive sport. They weren't hiding in Eastern European basements or operating from lawless territories. They were American high school students, living with their parents, attending classes by day and stealing millions by night. Their motivations mixed greed with glory, financial gain with social status in their underground community.

Joel Ortiz, considered one of the first to perfect the technique, began SIM swapping at seventeen. By twenty, he had stolen over five million dollars and was serving a ten-year prison sentence—the first person in America convicted for SIM swap fraud. His methodology became the template for hundreds of imitators. He would attend cryptocurrency conferences, identify wealthy attendees, and SIM swap them while they were distracted by presentations. One victim lost over a million dollars while giving a talk about blockchain security.

The social dynamics within SIM swapping crews resembled those of street gangs translated to digital space. Members earned respect through successful hits, with screenshots of cryptocurrency wallets serving as proof of prowess. They developed their own slang: "hitting" for attacking, "sauce" for methods, and "fullz" for complete identity profiles. They competed to hit the biggest targets, sometimes attacking the same victim simultaneously, racing to empty accounts.

The teenagers flaunted their wealth with a brazenness that defied logic. They posted Instagram photos of luxury cars, designer clothes, and stacks of cash. They flew first class to meet other members of their crew. They bought virtual items in video games for thousands of dollars. One member famously spent $100,000 on a single night at a club, paying with Bitcoin converted to cash that morning.

This ostentation became their downfall. Nicholas Truglia, arrested for stealing millions through SIM swaps, was caught partly because he couldn't stop bragging. He posted photos of himself with stolen money, discussed his crimes in Discord channels, and even livestreamed himself conducting attacks. When police arrested him, he was wearing a $500,000 watch purchased with stolen cryptocurrency.

The carrier employees became both unwitting accomplices and active participants. Some were simply fooled by sophisticated social engineering. Others, facing financial pressure and offered bribes that exceeded their annual salaries, actively facilitated swaps. Investigations revealed carrier employees selling SIM swaps for as little as $100, unaware or uncaring that they were enabling millions in theft.

The case of AT&T employee Brianna Mills illustrated this insider threat. Mills, making $15 per hour at a call center, was approached by SIM swappers offering $100 per successful swap. Over six months, she facilitated dozens of swaps, earning roughly $10,000 while enabling millions in theft. When arrested, she claimed she didn't know what the swappers were doing with the access, a claim investigators found hard to believe given the widespread media coverage of SIM swap attacks.

The victims of SIM swapping weren't just wealthy cryptocurrency investors. Small business owners lost access to their company accounts. Elderly individuals had their retirement savings drained. Social media influencers had their accounts hijacked and held for ransom. Dating app users faced extortion with intimate photos. The attack's versatility made everyone with a phone number a potential target.

The psychological impact on victims extended beyond financial loss. Robert Ross, who lost $1 million in cryptocurrency, described the experience as "digital rape." Victims reported PTSD symptoms, paranoia about technology, and a fundamental loss of trust in systems they'd relied on. Some abandoned digital services entirely, reverting to cash and in-person transactions.

The legal system struggled to address SIM swapping. The crimes crossed multiple jurisdictions—the attacker might be in Florida, the victim in California, the carrier in Texas, and the cryptocurrency exchange in New York. Prosecutors had to explain cryptocurrency to juries who barely understood email. Sentences varied wildly, with some teenage attackers receiving probation while others faced decades in prison.

Law enforcement developed specialized units to combat SIM swapping, but they faced significant challenges. The REACT task force (Regional Enforcement Allied Computer Team) in Santa Clara, California, became a model for investigating these crimes. They developed relationships with carriers, created rapid response protocols, and pioneered techniques for tracing cryptocurrency. Yet for every arrest they made, new attackers emerged.

The carriers' response ranged from inadequate to counterproductive. They implemented additional security measures—security questions, port freezes, in-store verification requirements—but these often inconvenienced legitimate customers more than they deterred attackers. Customer service representatives, measured on call times and customer satisfaction, faced pressure to help callers quickly, creating tension between security and service.

T-Mobile faced particular criticism after repeated breaches exposed customer data that facilitated SIM swaps. A 2021 breach affecting over 50 million customers included exactly the type of information—names, dates of birth, Social Security numbers—that SIM swappers needed. The company paid millions in settlements but continued to experience security incidents.

The cryptocurrency industry implemented its own defenses. Exchanges began offering non-SMS two-factor authentication, with some removing SMS options entirely. Hardware wallets became standard for serious investors. Some platforms implemented withdrawal delays and multi-signature requirements. Yet the fundamental vulnerability remained: most services still offered phone-based account recovery.

The evolution of SIM swapping techniques kept pace with defenses. Attackers moved from social engineering to technical exploits, using SS7 vulnerabilities to intercept text messages without swapping SIMs. They compromised carrier systems directly, bypassing customer service entirely. They developed automated tools that could attempt hundreds of swaps simultaneously.

The international dimension complicated everything. American teenagers collaborated with hackers in Nigeria, the Philippines, and Eastern Europe. Stolen cryptocurrency flowed through exchanges in Malta, the Cayman Islands, and Singapore. Investigators needed cooperation from countries with different laws, languages, and levels of technical capability.

The case of IOTA cryptocurrency revealed the global nature of these crimes. Attackers used SIM swapping to steal over $11 million in IOTA tokens from users worldwide. The investigation involved law enforcement from the UK, Germany, Canada, and the United States. The primary perpetrator, arrested in Oxford, England, was a 36-year-old who coordinated teenage SIM swappers across multiple countries.

Prevention became a personal responsibility that many weren't equipped to handle. Security experts recommended using Google Voice numbers for authentication, but this required technical knowledge many lacked. They suggested hardware tokens, but these cost money and were easy to lose. They advised using password managers, but these themselves became targets for SIM swappers seeking master passwords.

Michael Terpin, the victim from our opening, didn't accept his loss quietly. He sued AT&T for $224 million, arguing the carrier's negligence enabled the theft. He won a $75 million judgment against Nicholas Truglia, though collecting from a twenty-one-year-old criminal proved impossible. His cases established legal precedents but offered little comfort to victims who lacked his resources to pursue litigation.

The democratization of SIM swapping through crime-as-a-service models made the threat universal. Underground forums offered "SIM swapping tutorials" for $50. Telegram channels sold "carrier insiders" contacts. Discord servers provided real-time support for ongoing attacks. The barrier to entry dropped so low that any teenager with basic computer skills and criminal intent could attempt SIM swapping.

Today, SIM swapping remains a active threat despite increased awareness and security measures. The attacks have evolved from targeting individuals to focusing on companies, with attackers swapping corporate numbers to bypass business account protections. The rise of eSIMs introduced new attack vectors. The proliferation of financial apps that use phone numbers as primary identifiers created more valuable targets.

The fundamental tension remains unresolved: convenience versus security. Phone numbers were never designed to be security tokens, yet we've built an entire authentication infrastructure around them. Replacing this system would require coordinated action from thousands of companies and millions of users. Until then, our phones remain both our most trusted devices and our greatest vulnerabilities.

Lessons for Security Professionals

1. Implementing SIM Swap Resistant Authentication
- Remove SMS-based 2FA options entirely for high-value accounts
- Implement FIDO2/WebAuthn standards for passwordless authentication
- Use time-based one-time passwords (TOTP) via authenticator apps
- Deploy hardware security keys for privileged accounts
- Create account recovery processes that don't rely on phone numbers

2. Detection and Response Protocols
- Monitor for sudden authentication failures across multiple users
- Implement alerts for phone number changes on accounts
- Create honeypot accounts to detect targeted attacks early
- Develop rapid response procedures for suspected SIM swap attacks
- Establish direct relationships with carrier security teams

3. User Education and Training
- Conduct regular training on SIM swap risks and prevention
- Simulate SIM swap attacks as part of security awareness programs
- Provide clear guidance on securing personal phone accounts
- Create easy-to-follow incident response guides for victims
- Emphasize the importance of carrier account security

4. Architectural Security Improvements
- Implement additional verification for high-value transactions
- Create time delays for sensitive operations after authentication changes
- Use risk-based authentication that considers multiple factors
- Separate authentication methods for different privilege levels
- Design systems that gracefully handle authentication method compromises

5. Vendor and Carrier Management
- Pressure carriers to implement stronger swap protections
- Advocate for industry-wide standards on number porting security
- Require vendors to support non-SMS authentication methods
- Evaluate and document carrier security practices
- Develop contingency plans for carrier-level compromises

Lessons for Everyone

1. Protecting Your Phone Number
- Add carrier security features like port freezes or transfer PINs
- Use unique, complex passwords for carrier accounts
- Never share personal information that could be used to impersonate you
- Be cautious about where you publicly display your phone number
- Consider using separate numbers for authentication and communication

2. Reducing SIM Swap Impact
- Use authenticator apps instead of SMS for two-factor authentication
- Keep cryptocurrency in hardware wallets, not exchange accounts

- Maintain separate email accounts for financial services
- Document all your accounts and authentication methods
- Have backup access methods for critical accounts

3. Recognizing a SIM Swap Attack

- Sudden loss of cell service is the primary warning sign
- Unexpected "Welcome to carrier" messages indicate a swap
- Password reset emails you didn't request suggest ongoing attack
- Missing calls or texts that others say they sent
- Notifications about account changes you didn't make

4. Immediate Response Steps

- Contact your carrier immediately to report the swap
- Check all financial and cryptocurrency accounts
- Change passwords for critical accounts from a secure device
- Contact banks and exchanges to freeze accounts
- Document everything for law enforcement and insurance
- Alert friends and family who might receive scam messages

5. Long-term Protection Strategies

- Consider using Google Voice or VoIP numbers for authentication
- Invest in hardware security keys for important accounts
- Maintain offline backups of important data and credentials
- Regularly review and update account security settings
- Stay informed about evolving threats and protection methods
- Consider identity theft protection services

Part-I: Closing Reflection

"The moral dilemma of the computer age is that we have created a global nervous system before we have developed a global conscience. We are all interconnected, which means a vulnerability in one heart monitor is a vulnerability in our collective security."
— **William Stalling,** *Author of Computer Security: Principles and Practice*

The three chapters in Part I reveal a sobering truth about cybersecurity in the 21st century: our greatest vulnerability isn't in our code or our configurations—it's in ourselves. From Kevin Mitnick's pioneering social engineering to teenage financial hackers to SIM swap scammers, each story demonstrates how human trust, helpful nature, and technological dependence can be weaponized against us.

These aren't stories of sophisticated nation-states deploying zero-day exploits or advanced persistent threats using military-grade malware. These are stories of attackers who understood a fundamental principle: why break down the door when you can convince someone to open it for you?

The evolution from Mitnick's phone-based social engineering to modern SIM swapping shows how the core vulnerability—human trust—remains constant while the attack vectors evolve with technology. Each generation of technology introduces new ways to exploit the same human weaknesses. As we move toward an increasingly connected future with IoT devices, biometric authentication, and AI assistants, we must remember that the human factor will remain the critical vulnerability.

The age of many attackers forces us to reconsider our assumptions about threats. The teenager in their bedroom, armed with curiosity and time, can pose as significant a threat as organized crime syndicates. This democratization of attack capabilities means that everyone—from individuals to corporations to governments—must take security seriously.

Yet these stories also offer hope. Kevin Mitnick transformed from hacker to security advocate. Some teenage hackers have become valuable security researchers. Even SIM swapping has prompted improvements in authentication systems. Each attack teaches us lessons that, if properly applied, make us more resilient.

The challenge moving forward is not just technical but cultural. We must build security awareness into our education systems, teaching children not just to code but to code responsibly. We must create legal frameworks that balance punishment with rehabilitation, recognizing that young attackers might become our best defenders. We must design systems that are secure by default, not relying on users to make perfect security decisions.

As we prepare to explore the world of dark markets, hacktivism, and insider threats in Part II, remember that the vulnerabilities exposed in Part I underlie all cyber attacks. Whether the attacker is a nation-state or a teenager, a criminal syndicate or a hacktivist collective, they all exploit the same fundamental weakness: the human beings who design, operate, and use our digital systems.

The question isn't whether we can eliminate human vulnerability—we can't. The question is whether we can acknowledge it, account for it, and build defenses that protect us despite our inherent weaknesses. The stories in Part I suggest that we can, but only if we stop seeing security as a technical problem and start seeing it as a human one.

References and Notes

Chapter 1: The Human Factor
- Mitnick, K., & Simon, W. (2002). The Art of Deception: Controlling the Human Element of Security. Wiley.
- Shimomura, T., & Markoff, J. (1996). Takedown: The Pursuit and Capture of Kevin Mitnick. Hyperion.
- FitzGerald, M. (2014). "Kevin Mitnick, Once the World's Most Wanted Hacker, Is Now Selling Zero-Days to the Government." Motherboard/VICE.
- U.S. Department of Justice. (1999). "Kevin Mitnick Sentenced to Nearly Four Years in Prison." Press Release, August 9, 1999.
- Littman, J. (1996). The Fugitive Game: Online with Kevin Mitnick. Little, Brown and Company.

Chapter 2: The Teenage Bank Robber
- Australian Federal Police. (2019). "Teenager Charged with Cyber Attacks on Financial Institutions." Media Release, September 2019.
- Krebs, B. (2018). "The Rise of Juvenile Cybercrime." Krebs on Security.
- Financial Crimes Enforcement Network. (2020). "Advisory on Cybercrime and Cyber-Enabled Crime Exploiting the Coronavirus Disease Pandemic."
- Europol. (2019). "Internet Organised Crime Threat Assessment (IOCTA) 2019."
- National Crime Agency (UK). (2017). "Pathways into Cyber Crime Report."

Chapter 3: The SIM Swap Scammers
- Krebs, B. (2021). "The Wages of SIM-Swapping." Krebs on Security.
- U.S. Department of Justice. (2019). "Two Members of Infamous Hacking Group 'The Community' Sentenced for Scheme to Steal Cryptocurrency." Press Release, November 2019.
- Whittaker, Z. (2021). "T-Mobile Says Hackers Accessed Data of More Than 50 Million Current, Former and Prospective Customers." TechCrunch.
- REACT Task Force. (2020). "SIM Swapping: A Growing Threat to Digital Identity." White Paper.
- Terpin v. AT&T Mobility. (2020). Case No. 18-cv-06975. U.S. District Court, Central District of California.

Note: Some names and identifying details have been changed to protect individual privacy. These accounts are based on publicly documented incidents and court records.

This book draws from extensive reporting in cybersecurity media, including coverage by Darknet Diaries podcast by Jack Rhysider, Brian Krebs' security journalism, and investigative reporting by major news outlets. The author acknowledges the important work of security researchers and journalists who have documented these critical incidents for public awareness.

Part-II

Markets, Leaks, and Movements

The democratization of the internet brought unprecedented freedom—freedom to communicate, to commerce, and to challenge authority. But with this freedom came those who would exploit it for profit, politics, and power. Part II explores the dark markets that turned the internet into a criminal bazaar, the hacktivist movements that weaponized technology for ideology, and the insiders who shattered our assumptions about trust and secrecy.

These stories reveal how the same technologies that enable legitimate commerce, free expression, and government transparency also enable drug trafficking, vigilante justice, and devastating breaches of classified information. They force us to confront uncomfortable questions about privacy, freedom, and the price of security in a digital age.

From the libertarian dreams that birthed Silk Road to the anarchic chaos of Anonymous to the calculated betrayal of Edward Snowden, these chapters examine what happens when individuals decide that existing systems must be circumvented, exposed, or destroyed. Whether we view these actors as criminals or heroes—and many are seen as both—their actions have fundamentally shaped our digital world.

Chapter-4

Dark Markets - The Rise and Fall of Digital Black Markets

"Every time we close one dark market, two more appear. It's not a war we can win through enforcement alone."
— **Europol Director**, *after Operation Onymous*

October 1, 2013, San Francisco Public Library

Ross Ulbricht sat at his laptop in the science fiction section, fingers flying across the keyboard as he managed an empire worth hundreds of millions of dollars. To anyone watching, he looked like any other young programmer—scruffy beard, casual clothes, completely absorbed in his screen. The FBI agents who had been tracking him for months knew better. They knew that the unassuming 29-year-old was the Dread Pirate Roberts, creator and operator of Silk Road, the Amazon of illegal drugs.

Agent Gary Alford approached from behind while his partner, Thom Kiernan, came from the opposite direction. They had one chance. If Ulbricht closed his laptop, its encryption would make the evidence inaccessible. The entire investigation—years of work by dozens of agents—depended on the next few seconds.

A staged lover's quarrel erupted nearby. As Ulbricht turned to look at the commotion, Alford grabbed the laptop while Kiernan restrained the suspect. On the screen, still logged in, was the Silk Road administrator panel. The FBI had just captured the most wanted cybercriminal in America, along with the evidence needed to convict him.

The story of Silk Road began not with criminal intent but with ideological fervor. Ulbricht, a Penn State graduate with a master's in materials science, had become enamored with libertarian philosophy. He devoured the works of Ludwig vonO Mises and Murray Rothbard, convincing himself that drug prohibition was both immoral and ineffective. The government had no right, he believed, to tell people what they could put in their own bodies.

In 2010, Ulbricht discovered Bitcoin, the cryptocurrency that would make his vision possible. Here was a currency that existed outside government control, that could be transferred anonymously, that couldn't be seized or frozen. Combined with Tor, the anonymity network originally developed by the U.S. Navy, Bitcoin provided the technical foundation for a marketplace beyond the reach of law enforcement.

Silk Road launched in February 2011 with Ulbricht selling his own homegrown psychedelic mushrooms. The site's design was deliberately familiar—product listings with photos, vendor profiles with ratings, customer reviews, even a shopping cart. But instead of books or electronics, the products were heroin, cocaine, LSD, and hundreds of other illegal substances.

The escrow system Ulbricht implemented was brilliant in its simplicity. Buyers sent Bitcoin to

---

Silk Road, which held it until they confirmed receipt of their drugs. Only then would the site release payment to vendors, taking a commission of 6-10%. This eliminated the trust problem inherent in illegal transactions. Buyers couldn't be scammed by vendors who didn't deliver. Vendors couldn't be robbed by buyers who didn't pay.

The site's growth exceeded Ulbricht's wildest expectations. Within months, hundreds of vendors were offering thousands of products to tens of thousands of users. The forums buzzed with activity—discussions of drug quality, vendor reliability, even harm reduction advice. A strange sort of community formed, united by their participation in this illegal marketplace.

Ulbricht, posting as Dread Pirate Roberts (a reference to "The Princess Bride"), became a cult figure. His manifestos about freedom and non-aggression resonated with users who saw themselves not as criminals but as pioneers of a new economic order. He positioned Silk Road as a form of civil disobedience, a peaceful protest against unjust laws.

But maintaining this ideology became increasingly difficult as the site grew. Weapons appeared for sale, challenging Ulbricht's non-violence principles. Vendors scammed buyers despite the escrow system. Competitors emerged, trying to steal market share. Most troublingly, people claiming to be hackers threatened to release user information unless paid.

The transformation of Ross Ulbricht into Dread Pirate Roberts revealed the corrupting nature of power. Faced with threats to his empire, he allegedly conspired to have six people murdered, though none of these murders actually occurred. The peaceful libertarian who started Silk Road to reduce drug-related violence was now ordering hits like a crime boss.

Law enforcement's pursuit of Silk Road required unprecedented coordination and innovation. The DEA, FBI, IRS, Homeland Security, and even the Postal Service collaborated in joint task forces. They couldn't simply follow the money—Bitcoin transactions were pseudonymous. They couldn't trace internet connections—Tor obscured them. Traditional investigative techniques were useless.

The breakthrough came from careful analysis of the early days of Silk Road. IRS agent Gary Alford found posts on forums from early 2011 where someone using the name "altoid" promoted a new hidden service. The same username later posted looking for a Bitcoin programmer, including an email address: rossulbricht@gmail.com. This tiny mistake, made before Silk Road became notorious, would prove fatal.

Meanwhile, the FBI had located a Silk Road server in Iceland through traffic analysis. They imaged the server, finding a wealth of evidence including Bitcoin wallets, transaction records, and private messages. But they needed to connect this digital evidence to a real person. The altoid posts provided that connection.

The arrest at the San Francisco library was carefully orchestrated. Agents knew Ulbricht used public WiFi to avoid detection. They knew he carried his laptop everywhere, likely configured to encrypt everything if closed. The staged distraction—a couple loudly arguing—drew his attention just long enough for agents to grab the laptop while it was unlocked.

What they found exceeded their expectations. Not only was Ulbricht logged into Silk Road's administration panel, but his laptop contained journals detailing the site's creation, spreadsheets tracking its growth, and even his own writings about the alleged murder conspiracies. He had documented his own crimes in meticulous detail.

The trial of Ross Ulbricht became a landmark case in cryptocurrency and internet law. Prosecutors painted him as a drug kingpin who facilitated millions of transactions worth over a billion dollars. The defense argued he was a fall guy, that he had created Silk Road but handed it off to others who were the real Dread Pirate Roberts.

The jury didn't buy it. Ulbricht was convicted on all counts and sentenced to life in prison without parole—a sentence many considered excessive for someone who never physically touched drugs or directly harmed anyone. The judge justified the harsh sentence as necessary to deter others from creating similar marketplaces.

But Silk Road's closure didn't end dark markets—it accelerated their evolution. Within weeks of Ulbricht's arrest, Silk Road 2.0 launched, operated by former administrators of the original site. When it too was shut down, Silk Road 3.0 appeared. The brand had become bigger than any individual operator.

More sophisticated markets emerged. AlphaBay, launched in 2014, quickly surpassed Silk Road's size. It offered not just drugs but stolen data, hacking tools, and fraudulent documents. Its operator, Alexandre Cazes, learned from Ulbricht's mistakes. He lived in Thailand, beyond easy reach of U.S. law enforcement. He used multiple identities and complex operational security.

Yet Cazes made his own fatal error. He included his personal email address—pimp_alex_91@hotmail.com—in early AlphaBay password recovery emails. This mistake, discovered through painstaking analysis of thousands of transactions, led to his identification. When Thai police, working with the FBI, arrested him in July 2017, AlphaBay was processing over $600,000 in transactions daily.

Cazes never made it to trial. A week after his arrest, he was found dead in his Thai jail cell, an apparent suicide. AlphaBay died with him, but dozens of other markets continued operating. Hansa, Dream Market, Wall Street Market—each claimed to be more secure, more reliable, more trustworthy than the last.

The cat-and-mouse game between law enforcement and dark market operators became increasingly sophisticated. Operation Onymous in 2014 took down over 400 hidden services. Operation Bayonet in 2017 saw law enforcement secretly operating Hansa Market for a month, collecting information on vendors and buyers. Operation Dark Gold targeted vendors across multiple markets simultaneously.

Yet for every market closed, new ones opened. The hydra effect was in full display—cut off one head, and two more appeared. Markets implemented new security features: multi-signature transactions that prevented exit scams, decentralized architectures that had no single point of failure, even dead man's switches that would release user data if operators were arrested.

The economics of dark markets evolved beyond simple drug sales. Vendors offered subscription services for stolen streaming accounts. Hackers sold access to compromised corporate networks. Money launderers offered to clean cryptocurrency for a fee. The dark market ecosystem became a full-service criminal economy.

The human cost of dark markets extended far beyond their immediate participants. Fentanyl sold on these platforms fueled the opioid crisis, killing thousands. Stolen data enabled identity theft affecting millions. Hacking tools sold on markets were used in ransomware attacks on hospitals and schools. The libertarian dream of voluntary transactions between consenting adults had morphed into a nightmare enabling genuine harm.

Yet some argued dark markets actually reduced harm. Vendors had reputational incentives to sell pure products. Buyers could research substances and dosages in forums. The violence associated with street drug dealing was eliminated. These harm reduction arguments, while controversial, highlighted the complex moral landscape of dark markets.

The technology enabling dark markets continued advancing. Cryptocurrencies beyond Bitcoin—Monero, Zcash—offered enhanced privacy. Decentralized markets using blockchain technology eliminated central operators entirely. OpenBazaar attempted to create a legitimate peer-to-peer marketplace using the same technologies.

Law enforcement adapted too. Blockchain analysis companies like Chainalysis developed sophisticated tools for tracing cryptocurrency. International cooperation improved, with joint operations spanning dozens of countries. Undercover agents became sophisticated enough to operate as trusted vendors for months.

The psychological profile of dark market operators proved fascinating. Many weren't traditional criminals but technologically sophisticated individuals who saw themselves as entrepreneurs. They spoke of market share, customer service, and innovation. They competed on reliability and product quality. They had HR problems with staff and disputes with suppliers.

The story of Paul Le Roux illustrated this entrepreneurial criminality taken to extremes. Before becoming a cartel boss and DEA informant, Le Roux created E4M, a prescription drug network that pioneered many techniques later used by dark markets. His evolution from programmer to international criminal mastermind showed how technical skills combined with criminal ambition could create new forms of organized crime.

Today, dark markets remain a thriving ecosystem despite continuous law enforcement pressure. They've evolved from simple drug marketplaces to sophisticated platforms offering everything from fake vaccination cards to murder for hire (though the latter are almost always scams). The total volume is impossible to calculate but certainly reaches billions annually.

The legacy of Silk Road extends beyond criminal markets. It demonstrated that censorship-resistant, anonymous commerce was technically possible. This has implications for dissidents in authoritarian countries, whistleblowers exposing corruption, and anyone seeking financial privacy. The same technologies enabling drug sales also enable legitimate privacy-preserving commerce.

The philosophical questions raised by Silk Road remain unresolved. Should people have the right to buy and sell anything, even harmful substances? Can markets self-regulate through reputation systems? Is the violence of drug prohibition worse than the harm of drug use? These debates, started in Silk Road's forums, continue in policy discussions worldwide.

Ross Ulbricht remains in federal prison, serving a life sentence without parole. He's become a cause célèbre for some libertarians and criminal justice reform advocates who see his sentence as disproportionate. Others view him as a cautionary tale about the dangers of ideological extremism and the corrupting nature of power.

The dark market ecosystem he created continues evolving, adapting, surviving. Every closure spawns innovations. Every arrest teaches operational security lessons. Every seizure drives development of new technologies. The game of cat and mouse between law enforcement and dark market operators shows no signs of ending.

Lessons for Security Professionals

1. Understanding Underground Economics
- Study dark market operations to understand cybercriminal motivations and methods
- Monitor emerging markets for new threats and attack tools
- Track cryptocurrency flows to identify criminal networks
- Analyze vendor offerings to anticipate attack trends
- Understand the criminal supply chain from tools to services

2. Cryptocurrency Investigation Techniques
- Master blockchain analysis tools and techniques
- Understand privacy coins and mixing services
- Develop relationships with cryptocurrency exchanges
- Learn to trace funds across multiple blockchains
- Build cases despite pseudonymous transactions

3. Tor and Dark Web Operations
- Understand Tor's capabilities and limitations
- Develop techniques for de-anonymization when legally appropriate
- Create believable personas for undercover operations
- Monitor hidden services for threat intelligence
- Balance privacy rights with investigative needs

4. International Cooperation Requirements

- Build relationships with international law enforcement partners
- Understand legal frameworks in different jurisdictions
- Develop protocols for evidence sharing across borders
- Navigate different cultural approaches to cybercrime
- Create sustainable cooperation mechanisms

5. Disruption vs. Displacement

- Recognize that closures often displace rather than eliminate activity
- Plan for market migration when conducting operations
- Focus on systemic vulnerabilities rather than individual markets
- Develop strategies that increase criminal costs
- Measure success beyond simple closure metrics

Lessons for Everyone

1. Recognizing Dark Market Risks

- Understand that dark market purchases carry legal, health, and financial risks
- Know that law enforcement monitors these markets and arrests buyers
- Recognize that products may be contaminated, mislabeled, or fake
- Be aware that vendors may be law enforcement or scammers
- Understand that cryptocurrency transactions are traceable

2. Protecting Against Data Sales

- Monitor for your information being sold on dark markets
- Use identity monitoring services to detect compromised data
- Understand what information is valuable to criminals
- Take proactive steps to protect sensitive data
- Know how to respond if your data appears on markets

3. Understanding the Ecosystem

- Recognize how stolen data moves through criminal markets
- Understand the connection between data breaches and dark markets
- Know how ransomware groups use markets for tools and services
- Be aware of how personal information becomes weaponized
- Understand the underground economy's impact on cybersecurity

4. Avoiding Scams

- Be skeptical of services claiming to access dark markets safely
- Never trust "dark web scan" services that charge fees
- Understand that most sensational dark web claims are myths
- Know that "hiring a hacker" services are almost always scams
- Be wary of anyone claiming special dark web knowledge

5. Legal and Ethical Considerations

- Understand that accessing dark markets may be illegal
- Know that curiosity isn't a legal defense
- Recognize the real harm caused by dark market activities
- Consider the ethical implications of any involvement
- Support legitimate privacy tools while opposing criminal use

Chapter-5

Inside Anonymous - The Hacktivist Phenomenon

"Anonymous is not a group. It's an idea. And ideas are bulletproof."
— **Gabriella Coleman,** *anthropologist studying Anonymous*

December 10, 2010, London, England

Sixteen-year-old Jake Davis sat in his bedroom, watching history unfold on his computer screen. Across multiple IRC channels, hundreds of participants coordinated what would become the largest distributed denial-of-service attack in history. Their target: PayPal, Visa, and Mastercard—the financial giants who had blocked donations to WikiLeaks. Their weapon: a simple program called LOIC (Low Orbit Ion Cannon) that anyone could download and run. Their identity: Anonymous.

Davis, who would later gain infamy as "Topiary," the voice of the hacking group LulzSec, typed furiously in the coordination channels. "TANGO DOWN," someone announced as PayPal's website went offline. The chatroom erupted in celebration. They were David, slinging digital stones at corporate Goliaths. They were legion. They were Anonymous.

The operation, dubbed "Operation Payback," marked Anonymous's transformation from internet pranksters to a global force. What began in the anarchic culture of 4chan's /b/ board had evolved into something unprecedented: a leaderless collective capable of coordinated global action, united not by structure but by shared ideals and a distinctive aesthetic.

Anonymous's origins trace back to 2003, emerging from the primordial soup of 4chan, the imageboard where users posted anonymously by default. The name itself came from the default attribution—"Anonymous"—given to unsigned posts. This anonymity created a unique culture: irreverent, creative, cruel, and completely unpredictable. Users began to conceive of Anonymous as an entity, a hive mind formed from the collective actions of individual anons.

The first coordinated actions were purely for entertainment—"raids" on online games, prank calls to radio shows, and the systematic trolling of anyone who drew their attention. The targets were chosen almost randomly, their suffering documented and celebrated. It was cruel, juvenile, and utterly without political consciousness. Anonymous was, in their own words, doing it "for the lulz."

The transformation began in 2008 with Project Chanology, the campaign against the Church of Scientology. When the church tried to suppress a video of Tom Cruise making bizarre statements about Scientology, Anonymous responded with unexpected sophistication. They organized global protests, created professional-looking videos, and developed a coherent message about free speech and religious exploitation

---

The Guy Fawkes mask, popularized by the film "V for Vendetta," became Anonymous's symbol almost by accident. Protesters needed to hide their identities from Scientology's notorious legal team, and the mask was readily available, having been mass-produced for the recent film. The imagery was perfect: a smiling face hiding unknown intentions, a historical rebel commercialized into plastic, anonymity with a smirk.

February 10, 2008, saw something unprecedented: thousands of masked protesters appearing simultaneously at Scientology centers worldwide. They held signs with internet memes, played music, and handed out cake (the phrase "the cake is a lie" being a popular meme). It was playful yet serious, silly yet effective. The media didn't know what to make of it. Neither did law enforcement.

The success of Chanology attracted new participants who saw Anonymous not as trolls but as activists. These "moralfags," as the original anons derisively called them, wanted to use Anonymous's tactics for social justice. The tension between those who wanted to maintain the anarchic spirit and those who wanted to harness it for causes would define Anonymous's evolution.

Operation Payback began in September 2010 as retaliation against companies opposing piracy. When the Motion Picture Association of America hired companies to DDoS torrent sites, Anonymous DDoSed them back. The operations were crude but effective, taking down websites through sheer volume of traffic.

The WikiLeaks support operations in December 2010 changed everything. When financial companies blocked donations to WikiLeaks after it released U.S. diplomatic cables, Anonymous saw an attack on free speech. The response was massive. Thousands downloaded LOIC and pointed it at PayPal, Visa, and Mastercard. The sites crumbled under the assault.

But Anonymous was evolving beyond simple DDoS attacks. Hackers within the collective began conducting sophisticated intrusions. They dumped email databases, leaked documents, and defaced websites. The slogan "We do not forgive, we do not forget, expect us" became a chilling warning to those who drew their ire.

The Arab Spring of 2011 gave Anonymous a global purpose. Operations in Tunisia began when anons learned about the regime's censorship and oppression. They took down government websites, leaked censorship documents, and created care packages with tools to circumvent internet blocks. When Egyptians rose against Mubarak, Anonymous was there, attacking government sites and helping protesters maintain internet access.

The formation of LulzSec in May 2011 represented a splinter group taking Anonymous's methods to new extremes. Led by Hector Monsegur (Sabu), the group included Jake Davis (Topiary), Ryan Ackroyd (Kayla), and other who conducted high-profile attacks for entertainment. They hacked PBS in retaliation for a WikiLeaks documentary, stole user data from Sony, and leaked passwords from various sites.

LulzSec's fifty-day rampage demonstrated both the power and weakness of Anonymous-style operations. They gained massive media attention and caused significant damage. But they also drew intense law enforcement scrutiny. When Monsegur was arrested and turned informant, he helped authorities identify and arrest his confederates.

The arrests sent shockwaves through Anonymous. Jake Davis, the witty voice of LulzSec, turned out to be a teenager from the Shetland Islands. Ryan Ackroyd, who portrayed himself as a young woman named Kayla, was a British Army veteran. The revelations that trusted members were not who they claimed—or were actively working with law enforcement—created paranoia that persists today.

Yet Anonymous continued. Operations supported Occupy Wall Street, opposed SOPA/PIPA internet legislation, and targeted organizations from drug cartels to child pornography sites. The collective's actions ranged from noble to nihilistic, sometimes simultaneously. They were heroes to some, terrorists to others, and something incomprehensible to most.

The psychology of Anonymous participation proved complex. Members ranged from skilled hackers to people who simply ran LOIC during operations. Some participated in single operations while others devoted years to the collective. The lack of formal membership meant anyone could claim to be Anonymous, and everyone was simultaneously correct and incorrect.

The masked protests became a global phenomenon. From Turkey to Brazil, the Guy Fawkes mask appeared wherever people challenged authority. The mask transcended Anonymous, becoming a symbol of resistance. Time Warner, which owned the rights to the mask design, profited from every sale—a irony that perfectly captured the contradictions of anti-establishment movements in capitalist societies.

Law enforcement struggled to combat Anonymous. How do you arrest an idea? How do you infiltrate a group with no leadership, no membership lists, no formal structure? The FBI, NSA, and international agencies devoted enormous resources to tracking Anonymous operations, often catching individuals but never stopping the collective itself.

The tactics evolved with technology. Anonymous moved from IRC to encrypted channels. They adopted better operational security after watching colleagues arrested. They developed sophisticated technical capabilities while maintaining the original chaotic culture. They became, in essence, a weaponized meme.

The influence of Anonymous on global events remains debated. Did they help topple dictatorships or simply add noise to existing movements? Did they advance free speech or normalize vigilante justice? Were they freedom fighters or cyber terrorists? The answer depends entirely on perspective.

Gabriella Coleman, an anthropologist who embedded with Anonymous, described them as "weapons of the geek." They represented a new form of civil disobedience native to the internet age. Their actions forced conversations about power, privacy, and protest in digital spaces. They showed that anyone with a computer could challenge institutions previously thought untouchable.

The fragmentation of Anonymous was perhaps inevitable. Without structure, the collective couldn't maintain coherent goals. Different factions pursued contradictory operations. The banner of Anonymous was claimed by everyone from legitimate protesters to criminal extortionists. The lack of leadership that enabled its growth also prevented its evolution.

By 2016, Anonymous had largely faded from headlines, though operations continued at a smaller scale. The energy that drove Anonymous dispersed into other movements—the alt-right claimed some, social justice movements absorbed others. The tactics pioneered by Anonymous became standard for online activism across the political spectrum.

Yet Anonymous never truly died. The Twitter accounts still tweet. Operations still launch. The mask still appears at protests. Like a dormant virus, Anonymous exists in potential, ready to reactivate when conditions align. Every few years, events spark renewed activity, reminding the world that the collective endures.

The legacy of Anonymous extends beyond specific operations. They demonstrated that decentralized groups could achieve coordinated action without traditional organization. They proved that anonymity could be a source of power rather than weakness. They showed that the internet wasn't just a tool for existing power structures but a space where new forms of resistance could emerge.

The ethical questions raised by Anonymous remain unresolved. Is vigilante justice acceptable when legal systems fail? Can ends justify means when those means include destroying innocent people's data? Who decides what causes justify action? These questions become more urgent as Anonymous's tactics spread to other groups with different ideologies.

Today, the spirit of Anonymous lives on in various forms. Distributed Denial of Secrets publishes leaked documents. Hacktivists target authoritarian regimes. Protesters worldwide wear Guy Fawkes masks. The idea that anyone can challenge power through collective action persists, even as the original collective fragments.

Lessons for Security Professionals

1. Defending Against Ideological Threats
- Understand that ideological attackers have different motivations than criminals
- Recognize that public perception matters in responding to hacktivism
- Prepare for attacks during politically sensitive events
- Develop communication strategies for hacktivist incidents
- Balance security measures with avoiding appearing oppressive

2. DDoS Mitigation Strategies
- Implement robust DDoS protection including CDNs and scrubbing service

- Prepare for application-layer attacks, not just volumetric
- Develop incident response plans specific to DDoS scenarios
- Understand the legal implications of various defensive measures
- Test DDoS defenses regularly with simulated attacks

3. Managing Insider Threats from Sympathizers
- Recognize that employees may sympathize with hacktivist causes
- Monitor for unauthorized data access during sensitive periods
- Create clear policies about political activities and company resources
- Develop anonymous reporting mechanisms for concerns
- Balance security with employee rights and morale

4. Public Relations During Attacks
- Prepare crisis communication plans for hacktivist incidents
- Understand that hacktivists seek media attention
- Avoid responses that generate sympathy for attackers
- Coordinate legal and PR responses carefully
- Consider long-term reputational impacts of responses

5. Intelligence and Attribution Challenges
- Understand the difficulty of attributing anonymous attacks
- Develop intelligence sources in hacktivist communities
- Recognize false flag operations and misdirection
- Build relationships with researchers studying these groups
- Accept that some attacks may never be definitively attributed

Lessons for Everyone

1. Understanding Hacktivism
- Recognize that hacktivism affects regular users caught in crossfire
- Understand your data might be leaked for political reasons
- Know that being targeted doesn't mean you did anything wrong
- Be aware that anyone can claim to be Anonymous
- Distinguish between legitimate protest and criminal activity

2. Protecting Yourself During Operations
- Avoid participating in illegal activities like DDoS attacks
- Understand that using LOIC or similar tools is illegal
- Know that law enforcement monitors hacktivist channels
- Be skeptical of calls to action from unknown sources
- Recognize that anonymity tools don't guarantee legal protection

3. Data Protection
- Assume any online service could be targeted
- Use unique passwords to limit breach impact
- Monitor for your information in hacktivist leaks
- Understand that leaked data spreads permanently
- Take proactive steps to minimize exposed information

4. Critical Thinking About Hacktivism
- Evaluate hacktivist claims critically
- Understand that leaked documents may be selective or edited
- Recognize that hacktivists have their own agendas
- Consider unintended consequences of vigilante actions
- Support legitimate channels for addressing grievances

5. Personal Security Awareness
- Be cautious during politically charged events
- Understand that social media activity might draw attention
- Know that association with targets might affect you
- Prepare for service disruptions during major operations
- Have backup plans for critical online services

Chapter-6

The Insider Threat - When Trust Becomes Vulnerability

"I don't want to live in a world where everything I say, everything I do, everyone I talk to, every expression of creativity and love or friendship is recorded."
— **Edward Snowden,** *first public interview, June 2013*

June 6, 2013, Hong Kong

Edward Snowden sat in his room at the Mira Hotel, watching CNN International report on the NSA's PRISM program. The story, based on documents he had provided to journalists, was spreading across the globe. In a few days, he would reveal himself as the source, transforming from anonymous systems administrator to the most wanted man in the world. The thin, pale 29-year-old had just exposed the most extensive surveillance apparatus in human history.

The four laptops arrayed on his bed contained evidence of programs that sounded like science fiction: XKeyscore, which could search through virtually anyone's internet activity; Boundless Informant, tracking billions of phone calls and emails; Tempora, tapping into undersea fiber optic cables. Each revelation would shake the foundations of the intelligence community and spark a global debate about surveillance, privacy, and the price of security.

Snowden's path to becoming history's most consequential insider threat began unremarkably. A high school dropout who later earned a GED, he epitomized the self-taught technologist. His skills with computers compensated for his lack of formal education. The intelligence community, desperate for technical talent after 9/11, welcomed him with open arms.

His career trajectory through the intelligence community reads like a guided tour of America's secret world. He started as a security guard at the NSA's Center for Advanced Study of Language at the University of Maryland. Despite lacking a degree, his technical aptitude earned him rapid promotion. By 2006, he was working for the CIA in Geneva, managing network security for diplomats.

In Geneva, Snowden witnessed his first moral compromises. He watched CIA officers get a Swiss banker drunk and arrested for DUI, then offer to help in exchange for cooperation. He saw intelligence agencies spy on allies and enemies alike, with no distinction between threats and targets of opportunity. The idealistic young man who joined to serve his country began questioning what that service meant.

After leaving the CIA in 2009, Snowden worked as a contractor for Dell, managing NSA computer systems. This position gave him unprecedented access. As a systems administrator, he could see everything while remaining invisible. He had what the intelligence community calls "root" access—the keys to the kingdom. He could read anyone's emails, access any database, copy any file.

The contractors operating within the intelligence community represented a unique vulnerability. Unlike government employees who underwent regular polygraphs and psychological evaluations, contractors faced less scrutiny. They often had access equal to or exceeding regular employees but with less oversight. The intelligence community had outsourced much of its IT infrastructure, creating a shadow workforce with extraordinary privileges.

Snowden's growing disillusionment accelerated after Obama's election. He had hoped the new administration would curtail Bush-era surveillance programs. Instead, they expanded. The same president who campaigned on transparency presided over unprecedented surveillance growth. Snowden watched the NSA collect millions of Americans' communications without warrants, violating what he saw as constitutional principles.

The technical capabilities Snowden witnessed exceeded public imagination. The NSA had backdoors in major technology companies' servers. They could activate webcams and microphones remotely. They collected metadata on virtually every phone call in America. They had compromised encryption standards, weakening security for everyone. The agency charged with protecting American communications was systematically undermining them.

In March 2013, Snowden accepted a position with Booz Allen Hamilton in Hawaii, specifically to gather evidence of these programs. This premeditation would later fuel debates about his motivations. Was he a whistleblower documenting wrongdoing or a spy conducting reconnaissance? His methodical collection of documents suggested careful planning rather than spontaneous conscience.

The scope of Snowden's theft was staggering. He claimed to have accessed 1.7 million documents, though he said he carefully reviewed what he took. He used web crawlers to automatically download files, exploiting his administrator privileges to bypass security controls. He transferred files to encrypted drives, sometimes hiding them in a Rubik's Cube on his desk. His colleagues, trusting him as one of their own, never suspected.

The psychology of insider threats reveals common patterns that Snowden exemplified. He was intelligent but felt underappreciated. He had strong ideological beliefs that conflicted with his organization's actions. He possessed technical skills that exceeded his supervisors' understanding. He worked in a high-trust environment with minimal supervision. These factors created perfect conditions for betrayal.

Snowden's decision to flee to Hong Kong was strategic. The semi-autonomous region had extradition treaties with the U.S. but also a tradition of protecting dissidents. He calculated that China wouldn't immediately surrender him but also wouldn't want him permanently. He needed time to ensure the documents reached journalists and the story broke before authorities silenced him.

The journalists Snowden chose—Glenn Greenwald, Laura Poitras, and Barton Gellman—were selected carefully. Greenwald was a fierce critic of surveillance overreach. Poitras was already

---

under government scrutiny for her documentaries. Gellman had experience handling classified materials. Snowden didn't just leak; he orchestrated a coordinated media campaign.

The revelations rolled out strategically, each building on the last. First came PRISM, showing tech company cooperation with surveillance. Then phone metadata collection, affecting every American. Then international surveillance, angering allies. Then offensive cyber operations, revealing America's digital weapons. Each story generated headlines before the next dropped, maintaining momentum.

The government's response was swift but ineffective. They revoked Snowden's passport while he was traveling to Moscow, stranding him in Russia—hardly the outcome they intended. They charged him under the Espionage Act, ensuring he couldn't return without facing life in prison. They pressured journalists and newspapers, sometimes successfully suppressing stories but often generating more publicity.

The damage to intelligence operations was immediate and severe. Terrorist groups changed communication methods. Foreign governments hardened their systems. Technology companies implemented encryption to prevent government access. Intelligence sharing between allies decreased. Programs that took years to develop were rendered useless overnight.

Yet Snowden's supporters argued the damage to democracy from secret surveillance exceeded any intelligence loss. They pointed to the NSA's own violations of law, documented in FISA court opinions. They noted that government officials had lied to Congress about surveillance scope. They argued that Snowden revealed not just capabilities but abuse.

The insider threat Snowden represented wasn't unique. Chelsea Manning had leaked diplomatic cables to WikiLeaks. Thomas Drake, William Binney, and other NSA whistleblowers had tried legal channels before going public. Reality Winner would later leak evidence of Russian election interference. Each case revealed systemic issues with handling dissent within intelligence agencies.

The reforms following Snowden's revelations were significant but contested. The USA Freedom Act ended bulk phone metadata collection. Technology companies strengthened encryption. The intelligence community implemented new insider threat programs. But many argued these changes were cosmetic, that surveillance had simply evolved rather than decreased.

The personal cost to Snowden was enormous. He lost his country, his freedom, his family connections. He lived in Moscow airport for 40 days before Russia granted asylum. He existed in legal limbo—unable to travel, dependent on Putin's protection, a chess piece in geopolitical games. His girlfriend joined him, but his life was constrained by constant surveillance and limited options.

The technology industry's response revealed the sector's complicated relationship with government. Companies that had quietly cooperated with surveillance suddenly championed privacy. They implemented end-to-end encryption, fought government requests, and marketed security features. Yet they continued working with intelligence agencies on other projects, maintaining the ambiguous relationship between Silicon Valley and Washington.

The international impact reshaped global internet governance. Countries questioned American control of internet infrastructure. Data localization laws proliferated as nations sought to protect citizens' information. The European Union strengthened privacy regulations. The dream of a unified global internet fragmented into regional networks with different rules and protections.

The psychological profile of insider threats evolved after Snowden. Organizations recognized that technical capability combined with ideological motivation created unique risks. They implemented behavioral monitoring, looking for signs of disgruntlement or unauthorized access. They restricted administrator privileges and implemented zero-trust architectures. They created insider threat programs that sometimes caught criminals but also created paranoid workplaces.

Today, Snowden remains in Moscow, his temporary asylum repeatedly extended. He works as a privacy advocate, speaking at conferences via video link, writing about surveillance, developing privacy tools. He's simultaneously revered as a hero and reviled as a traitor, his legacy dependent entirely on perspective.

The broader implications of the Snowden revelations continue unfolding. Every major breach now raises questions about insider threats. Organizations struggle to balance security with employee trust. Whistleblower protections remain weak, forcing potential revealer of wrongdoing to choose between silence and exile. The surveillance capabilities he revealed have only grown more sophisticated.

Lessons for Security Professionals

1. Comprehensive Insider Threat Programs
- Implement behavioral analytics to detect unusual access patterns
- Monitor privileged account usage continuously
- Create psychological profiles identifying potential risks
- Establish clear escalation procedures for suspicious activity
- Balance security measures with employee privacy and morale

2. Privileged Access Management
- Implement least privilege principles rigorously
- Require multi-person authorization for sensitive operations
- Rotate administrative credentials regularly
- Monitor and log all privileged actions
- Segment access to prevent single individuals from accessing everything

3. Contractor and Third-Party Risk
- Apply same security standards to contractors as employees
- Limit contractor access to essential systems only

- Regularly review and audit contractor permissions
- Implement time-based access that expires automatically
- Monitor contractor activity more closely than employees

4. Data Loss Prevention Strategies
- Deploy DLP tools to detect mass downloads
- Monitor for unauthorized encryption or compression
- Track removable media usage
- Implement network segmentation to limit access
- Create honeypots to detect unauthorized access

5. Creating Positive Security Culture
- Establish legitimate channels for reporting concerns
- Address employee grievances before they become threats
- Foster environment where security questions are encouraged
- Recognize that excessive security can create resentment
- Build trust while maintaining vigilance

Lessons for Everyone

1. Understanding Your Digital Footprint
- Assume digital communications may be monitored
- Understand metadata reveals patterns even without content
- Know that deleted doesn't mean gone
- Recognize that privacy policies may not protect from government access
- Consider using encryption for sensitive communications

2. Protecting Personal Privacy
- Use end-to-end encrypted messaging apps
- Enable two-factor authentication on all accounts
- Regularly review app permissions and access
- Understand the trade-offs between convenience and privacy
- Know your rights regarding data collection and surveillance

3. Recognizing Insider Threat Indicators
- Be aware of colleagues exhibiting unusual behavior
- Report suspicious activity through appropriate channels
- Understand that insider threats can be ideological, not just criminal
- Know that excessive curiosity about others' work may indicate problems
- Balance vigilance with avoiding paranoid workplace culture

4. Whistleblowing Considerations
- Understand legal protections and their limitations
- Know internal reporting channels and when they fail
- Recognize personal costs of whistleblowing
- Seek legal counsel before taking action
- Document everything if witnessing wrongdoing

5. Balancing Security and Privacy
- Understand that absolute security and absolute privacy are incompatible

- Make informed decisions about privacy trade-offs
- Support transparency in government surveillance programs
- Engage in democratic processes about surveillance policies
- Recognize that privacy is a fundamental right worth protecting

Part II: Closing Reflection

"Technology is a double-edged sword; the same encryption that protects our privacy protects a criminal's plans. We are in a high-stakes battle over how much freedom we will trade for security. In the digital age, information is power—and when you give an individual the power of a nation-state, you also give them the potential for a nation-state's worth of damage."
— **Bruce Schneier,** *Internationally renowned security technologist, called a "security guru" by The Economist.*

The three chapters in Part II demonstrate how internet access initially brought freedom to users but ultimately developed a dual nature that became dangerous. The three stories about Silk Road, Anonymous, and Edward Snowden illustrate what happens when individuals choose to challenge established systems of justice or freedom.

Ross Ulbricht built his libertarian marketplace through voluntary exchange principles, but allegedly ordered murders to defend his business empire. The group Anonymous began as a prank group focused on "lulz" before it evolved into a worldwide force that equally enjoyed destroying governments and killing people without reason. Edward Snowden revealed surveillance programs that he believed violated human rights, yet his actions led to damage to intelligence operations that protected human lives.

These narratives show more than basic moral lessons about good versus evil characters. The stories show how people who fought for freedom, justice, and truth ended up creating effects they could not control. The Silk Road platform decreased drug-related violence, but it made it possible for users to develop drug dependence. Through their actions, Anonymous protected vulnerable groups, but they also used their power to destroy innocent people for their own entertainment. The disclosure of unconstitutional surveillance by Snowden revealed government misconduct, but it threatened to compromise national security operations that protected human lives.

The main link between these stories emerges from the continuous battle between individual freedom and collective responsibility. The protagonists took illegal actions because they believed the existing legal framework operated as an unjust system. The individuals took on two positions by enforcing laws while making judicial decisions about which information to disclose, which businesses to sustain, and which secrets to reveal. Their actions created essential questions about democratic operations and citizen participation in deciding national issues.

The technological instruments that enabled these actions were designed to safeguard privacy rights and support individual liberty. The privacy protection tools and freedom-enabling tools have become instruments for criminal activities and destructive conduct. The fundamental principle of technology shows that freedom-enabling tools can generate harmful effects. Activists who use protective measures against surveillance become vulnerable to attacks from predators. The digital currency system, which provides users with financial freedom by circumventing the banking system, makes it possible for hackers to conduct ransomware attacks. The encryption system, which protects personal information, also protects criminal organizations from detection.

The responses to these new security risks have evolved into complex solutions. Law enforcement agencies created new tracking methods to identify users in anonymous networks and monitor digital currency transactions. Governments used the threats their actors created to build their power base while expanding their surveillance capabilities. Users must give up their privacy rights and freedom to access enhanced security measures that organizations have implemented. Security measures and countermeasures in an ongoing battle continue to intensify with each new technological advancement.

Human society depends on the Internet as its core foundation because it contains all social conflicts and opposing elements that exist in physical space. The digital utopia failed to deliver complete government independence because unregulated systems allowed influential individuals to dominate defenseless groups. The idea that information naturally seeks freedom has proven incorrect because specific details can become dangerous when they enter public access.

The stories demonstrate how individuals can harness their power to bring about positive transformations in their communities. Ross Ulbricht demonstrated through his criminal activities that alternative economic systems could function in reality. The decentralized structure of Anonymous allowed members to start vital discussions about digital justice and power systems. Edward Snowden revealed classified information, which triggered worldwide discussions about surveillance and privacy that resulted in significant changes to national security legislation.

Part-III

Cyber Weapons and Nation-States

Introduction to Part III

The militarization of cyberspace represents one of the most significant developments in modern warfare. Nation-states have transformed code into weapons, keyboards into launch platforms, and networks into battlefields. Part III explores this new domain of conflict where attacks move at the speed of light, attribution remains murky, and the distinction between war and peace blurs beyond recognition.

These stories reveal how governments have weaponized the same technologies that power our daily lives. From Russia's hybrid warfare combining hacking with propaganda to ransomware attacks that paralyze hospitals to surveillance tools that turn smartphones into spy devices, we examine the dark reality of state-sponsored cyber operations.

The stakes couldn't be higher. These aren't just attacks on computers—they're attacks on democracy, healthcare, human rights, and the foundations of civil society. As nation-states and their proxies wage silent wars through our networks, civilians find themselves on the front lines of conflicts they may not even know exist.

Chapter-7

The Nation-State Hacker - APT28 and State-Sponsored Espionage

"This wasn't just an attack on data or computers. This was an attack on democracy itself."
— **John Podesta,** *after the DNC hack*

March 19, 2016, Washington D.C.

John Podesta, Hillary Clinton's campaign chairman, received an email that appeared to be from Google. "Someone has your password," it warned, providing a link to reset his credentials. His assistant, following protocol, forwarded it to the campaign's IT staff for verification. The response came back: "This is a legitimate email." That single word—legitimate instead of illegitimate—would alter the course of American history.

The email wasn't from Google. It was from APT28, also known as Fancy Bear, a Russian military intelligence unit conducting one of the most consequential cyber operations ever undertaken. Within hours, they had access to Podesta's entire email archive—years of communications that would soon weaponize American democracy against itself.

APT28 represents the evolution of espionage in the digital age. Where Cold War spies needed dead drops and secret meetings, modern intelligence officers need only keyboards and internet connections. Where stealing documents once required physical access and cameras, now terabytes can be exfiltrated without leaving Moscow. The GRU Unit 26165, the force behind APT28, had transformed intelligence gathering from artisanal craft to industrial process.

The group's origins trace to the mid-2000s when Russia recognized cyberspace as a domain for strategic competition. The 2007 cyber attacks on Estonia, triggered by the relocation of a Soviet war memorial, demonstrated cyber operations' potential. While attribution for Estonia remained officially ambiguous, the attacks showed Russia that digital weapons could achieve political objectives without conventional military force.

APT28's targeting revealed Russian strategic priorities: NATO, former Soviet states, defense contractors, and democratic institutions. Their victims included the German Bundestag, the French television station TV5Monde, the World Anti-Doping Agency, and countless government agencies across Europe and America. Each operation gathered intelligence while also serving broader psychological and political objectives.

The technical sophistication of APT28 evolved continuously. Their malware families—Sofacy, X-Agent, X-Tunnel—demonstrated advanced capabilities. They used zero-day exploits when necessary but often succeeded through simpler means. Why burn valuable vulnerabilities when spear-phishing worked perfectly? Their infrastructure sprawled across compromised servers worldwide, creating layers of obfuscation that complicated attribution.

The 2016 targeting of American political organizations represented a paradigm shift in cyber operations. Previous state-sponsored intrusions sought intelligence to inform policy decisions. APT28 sought intelligence to influence electoral outcomes. They crossed the line from espionage to active measures, from gathering information to weaponizing it.

The Democratic National Committee breach began in summer 2015, but accelerated in March 2016 when APT28 gained access to the Democratic Congressional Campaign Committee network. From there, they moved laterally, eventually compromising over 30 DNC computers. They installed keystroke loggers, screenshot capture tools, and data exfiltration malware. For months, they watched everything, invisible observers of American democracy's inner workings.

The sophistication lay not in the breach but in the operation that followed. APT28 didn't just steal data—they strategically released it. They created personas like "Guccifer 2.0" claiming to be Romanian hackers. They provided documents to WikiLeaks, ensuring maximum impact. They timed releases for political effect, dropping batches when they would cause maximum damage.

The psychological warfare aspect of APT28's operations deserves special attention. They understood that information's value wasn't just its content but its context and timing. Emails showing normal political operations—internal debates, strategy discussions, personal opinions—became weapons when released selectively. They turned democracy's openness against itself.

Attribution of APT28 activities demonstrated both the possibilities and limitations of cyber forensics. Security firms like CrowdStrike and FireEye identified distinctive patterns: Russian-language settings in malware, operational hours matching Moscow business days, infrastructure reuse across operations. The Mueller investigation later indicted twelve GRU officers by name, providing unprecedented detail about their operations.

Yet attribution remained contested. Russia denied involvement, offering alternative theories. Some cybersecurity experts questioned aspects of the evidence. The ambiguity inherent in cyber operations—the ease of false flags, the difficulty of proving state direction—meant that even overwhelming evidence couldn't achieve universal acceptance. This attribution problem would become a defining challenge of cyber conflict.

The impact extended far beyond the immediate targets. Every political organization worldwide suddenly realized they were potential targets. Every election became a potential battlefield. The trust essential to democratic processes eroded as citizens questioned whether leaks were genuine, selective, or fabricated. APT28 had weaponized doubt itself.

The technical analysis of APT28's tools revealed an interesting paradox. While capable of sophisticated operations, they often chose simple, reliable methods. Their spear-phishing emails weren't particularly sophisticated—many contained grammatical errors or suspicious elements. But they sent thousands, knowing only one person needed to click. Quantity had a quality of its own.

The group's operational security varied wildly. Sometimes they demonstrated exceptional tradecraft, using encrypted channels and careful infrastructure management. Other times they made amateur mistakes, reusing email accounts or failing to use VPNs. This inconsistency suggested either multiple teams with varying skill levels or a calculation that perfect operational security wasn't necessary when operating from Russian territory.

The broader ecosystem of Russian cyber operations included groups beyond APT28. APT29 (Cozy Bear), associated with the SVR, conducted more traditional espionage. Sandworm, linked to the GRU, focused on destructive attacks. Criminal groups operated with apparent state tolerance, conducting ransomware attacks that aligned with Russian interests. The boundaries between state, criminal, and patriotic hackers blurred deliberately.

The international response to APT28's activities revealed the challenges of deterring state-sponsored cyber operations. Sanctions targeted individuals who were unlikely to ever leave Russia. Diplomatic expulsions were met with reciprocal measures. "Naming and shaming" had limited effect on a country that denied involvement. Traditional deterrence models, developed for nuclear weapons, proved inadequate for cyber conflicts.

The evolution of APT28's tactics continued after 2016. They targeted the 2018 Winter Olympics after Russia was banned for doping, deploying destructive malware disguised as ransomware. They breached the Organisation for the Prohibition of Chemical Weapons during investigations of Russian chemical weapon use. Each operation demonstrated cyber operations integrated with broader strategic objectives.

The defensive improvements prompted by APT28 were significant but insufficient. Organizations implemented better email filtering, security awareness training, and incident response procedures. But defending against nation-state actors with effectively unlimited resources, time, and motivation remained extraordinarily difficult. The asymmetry favored attackers who needed to succeed only once while defenders needed to succeed always.

Lessons for Security Professionals

1. Nation-State Threat Modeling
- Assess whether your organization could be a nation-state target
- Understand that targeting may be indirect (supply chain, partner organizations)
- Recognize that geopolitical events increase targeting likelihood
- Plan defenses assuming unlimited attacker resources
- Accept that prevention may be impossible; focus on detection and resilience

2. Advanced Email Security
- Implement DMARC, SPF, and DKIM authentication
- Deploy advanced threat protection scanning attachments and links
- Train employees specifically on spear-phishing recognition
- Create processes for verifying suspicious emails
- Implement banner warnings for external emails

3. Lateral Movement Prevention

- Segment networks to limit breach impact
- Implement privileged access management rigorously
- Monitor for unusual internal network activity
- Deploy deception technologies to detect intrusions
- Regularly hunt for threats; don't rely solely on alerts

4. Intelligence-Driven Defense

- Subscribe to threat intelligence feeds
- Participate in information sharing organizations
- Understand adversary TTPs (Tactics, Techniques, Procedures)
- Implement indicators of compromise from similar organizations
- Develop relationships with law enforcement and intelligence agencies

5. Incident Response for Nation-State Attacks

- Plan for long-term persistent threats
- Assume total compromise when nation-states are involved
- Coordinate with government agencies early
- Prepare for public relations challenges
- Document everything for potential legal/diplomatic responses

Lessons for Everyone

1. Email Vigilance

- Verify sender addresses character by character
- Never click links in unexpected emails
- Confirm password reset requests through separate channels
- Be especially cautious during political/sensitive periods
- Report suspicious emails to IT/security teams

2. Personal Operational Security

- Use unique, complex passwords for every account
- Enable two-factor authentication everywhere possible
- Be cautious about information shared on social media
- Understand that personal accounts may be targeted to reach employers
- Keep software updated to patch known vulnerabilities

3. Information Warfare Awareness

- Recognize that leaked information may be selective or edited
- Verify inflammatory claims through multiple sources
- Understand timing of leaks may be manipulative
- Be skeptical of anonymous sources claiming insider knowledge
- Consider cui bono (who benefits) from information releases

4. Democratic Resilience

- Maintain faith in democratic processes despite interference attempts
- Report suspected foreign influence operations
- Educate others about information warfare tactics

- Support transparency and security in electoral systems
- Participate in democracy despite attempts to discourage it

5. Organizational Security Culture

- Take security training seriously
- Report suspicious activity promptly
- Don't circumvent security measures for convenience
- Understand that everyone is a potential target
- Build collective defense through individual vigilance

Chapter-8

Ransomware Rampage - The WannaCry Global Crisis

"The attack was stoppable. A patch had been available for two months. This was a completely preventable disaster that we allowed to happen."
— **Brad Smith,** *Microsoft President, after WannaCry*

May 12, 2017, 8:00 AM, London

The first calls to the NHS IT helpdesk seemed routine—users reporting error messages on their computers. Within an hour, the calls became a flood. Across Britain's National Health Service, computer screens displayed the same terrifying message: "Oops, your files have been encrypted!" Beneath a red padlock icon, text demanded $300 in Bitcoin to decrypt files. WannaCry had arrived.

By noon, emergency rooms were diverting patients. Surgeries were being cancelled. Medical devices running Windows XP—devices that couldn't be updated without losing regulatory certification—were failing. Doctors reverted to pen and paper. Nurses couldn't access patient records. The NHS, already stretched thin, was in crisis. And it was just the beginning.

WannaCry represented the weaponization of a cyber weapon allegedly developed by the NSA. The EternalBlue exploit, targeting a vulnerability in Windows SMB protocol, had been stockpiled by intelligence agencies for surveillance purposes. When the Shadow Brokers—a mysterious group that remains unidentified—leaked NSA tools in April 2017, they released digital weapons into the wild. Within weeks, criminals had weaponized them.

The ransomware's propagation mechanism made it uniquely dangerous. Unlike typical ransomware that required user interaction—clicking a link, opening an attachment—WannaCry spread automatically. It scanned for vulnerable systems and infected them without any human involvement. It was a digital pandemic, spreading through networks like biological viruses spread through populations.

Patient zero remains disputed, but the infection's spread was meticulously documented. From its initial infection point, WannaCry spread to 150 countries within hours. It hit Telefónica in Spain, disrupting phone services. It infected Deutsche Bahn, causing train delays across Germany. It crippled Renault factories in France, halting production. It spread through Russia's Interior Ministry, infecting over 1,000 computers.

The technical analysis revealed both sophistication and amateurism. The ransomware component was relatively standard, using RSA-2048 encryption to lock files. The payment mechanism was poorly designed, using only three Bitcoin wallets for all victims, making it impossible to automatically verify payments. But the EternalBlue exploit was military-grade, capable of spreading through any unpatched Windows system.

Microsoft had actually released a patch for the vulnerability in March 2017, two months before WannaCry. But millions of systems remained unpatched. Organizations delayed updates fearing compatibility issues. Home users ignored update notifications. Systems running pirated Windows couldn't update. End-of-life systems like Windows XP had no patches available—until Microsoft took the extraordinary step of releasing emergency patches for unsupported systems.

The NHS's vulnerability highlighted systemic healthcare IT problems. Medical devices ran ancient operating systems because updating might void FDA approval. Hospital networks mixed critical life-support systems with administrative computers. IT budgets were consistently underfunded, with money prioritized for direct patient care. The technical debt accumulated over decades came due in a single morning.

At University Hospital of North Durham, Dr. Krishna Chinthapalli watched the chaos unfold. Operating theaters shut down. The pathology lab couldn't process blood tests. Radiology couldn't access previous images for comparison. "We had elderly patients with complex conditions," he later testified, "and we couldn't see their medical histories. We were flying blind."

The human impact extended beyond inconvenience. Ambulances were diverted, potentially delaying critical care. Cancer treatments were postponed. Surgeries were cancelled. While the NHS claimed no deaths directly resulted from WannaCry, the full impact on patient outcomes may never be known. How do you measure the effect of delayed diagnoses or postponed treatments?

The kill switch discovery by Marcus Hutchins seemed like providence. The 22-year-old security researcher, analyzing WannaCry's code, noticed it checked for a specific unregistered domain. Thinking he was setting up a sinkhole to track infections, he registered the domain for $10.69. Inadvertently, he had activated a kill switch that stopped WannaCry's spread.

The kill switch's existence raised intriguing questions. Was it intentional—a safeguard by WannaCry's creators? Or accidental—a debugging feature left in production code? Some speculated it was designed to prevent analysis in sandbox environments that simulate internet connectivity. Regardless, its activation saved potentially billions in additional damage.

But variants quickly emerged without kill switches. WannaCry 2.0, 3.0, and numerous copycats appeared within days. Security researchers played whack-a-mole, finding and triggering kill switches when they existed, developing decryption tools when possible, racing against attackers who learned from each iteration.

The attribution of WannaCry became a geopolitical issue. The United States, United Kingdom, and other nations eventually attributed it to North Korea, specifically the Lazarus Group. The evidence included code similarities to previous North Korean attacks, infrastructure overlap, and linguistic analysis. North Korea denied involvement, but the attribution aligned with their history of cybercrime for revenue generation.

The financial impact was staggering. Estimates ranged from hundreds of millions to billions in damages. FedEx's TNT Express subsidiary lost $400 million. Merck pharmaceutical reported $870 million in losses. Maersk shipping, infected through their Ukrainian subsidiary, lost $300 million despite WannaCry not being the primary malware in their case. The cyber insurance industry faced claims that tested policy limits and exclusions.

Yet WannaCry's creators earned relatively little—roughly $140,000 in Bitcoin payments. The poor payment infrastructure meant many victims who wanted to pay couldn't determine how. Others who paid never received decryption keys. The attack seemed more destructive than profitable, raising questions about whether ransomware was the true goal or merely camouflage for nation-state destruction.

The incident sparked global conversations about vulnerability disclosure. Should governments stockpile zero-days for offensive operations? The Vulnerabilities Equities Process, meant to balance offensive and defensive needs, clearly failed. EternalBlue, kept secret for years, enabled massive damage when leaked. Critics argued that governments should disclose vulnerabilities to vendors immediately, prioritizing global security over intelligence gathering.

Microsoft's Brad Smith delivered a scathing blog post comparing government vulnerability stockpiling to the military losing tomahawk missiles. He called for a "Digital Geneva Convention" limiting cyber weapons. The tech industry, long criticized for security failures, found itself able to deflect blame toward governments whose cyber weapons had been stolen and weaponized.

The healthcare sector underwent painful introspection. The FDA reconsidered policies preventing medical device updates. Hospitals invested in network segmentation and backup systems. Healthcare-specific security frameworks emerged. But the fundamental tension between innovation, regulation, and security remained unresolved.

Lessons for Security Professionals

1. Patch Management Excellence
- Implement automated patching where possible
- Develop risk-based patching prioritization
- Maintain accurate asset inventories
- Create emergency patching procedures
- Test patches in isolated environments first

2. Network Segmentation Strategy
- Isolate critical systems from general networks
- Implement east-west traffic inspection
- Use VLANs and firewalls to contain breaches
- Regularly test segmentation effectiveness
- Design networks assuming compromise will occur

3. Backup and Recovery Planning

- Maintain offline, immutable backups
- Test restoration procedures regularly
- Document recovery time objectives
- Store backups geographically distributed
- Protect backups from ransomware specifically

4. Crisis Management Preparation

- Develop ransomware-specific incident response plans
- Establish communication protocols during outages
- Prepare for manual operations if systems fail
- Create decision trees for payment considerations
- Coordinate with law enforcement preemptively

5. Threat Intelligence Integration

- Monitor for ransomware indicators continuously
- Subscribe to early warning systems
- Share intelligence with industry peers
- Understand ransomware groups' TTPs
- Track cryptocurrency wallets and infrastructure

Lessons for Everyone

1. Personal Ransomware Prevention

- Keep all software updated, especially operating systems
- Use reputable antivirus software
- Backup important files regularly
- Be cautious with email attachments
- Avoid pirated software that can't receive updates

2. Recognizing Ransomware Attacks

- Files suddenly becoming inaccessible
- Strange file extensions appearing
- Ransom notes on desktop or in folders
- Computer running slowly or unusually
- Antivirus software being disabled

3. Immediate Response Actions

- Disconnect from network immediately
- Power off to prevent spread
- Don't pay ransom—no guarantee of recovery
- Report to authorities
- Seek professional help for recovery

4. Backup Best Practices

- Follow 3-2-1 rule: 3 copies, 2 different media, 1 offsite
- Test backups regularly
- Keep some backups offline
- Version backups to recover from different points

- Encrypt backups for privacy

5. Organizational Awareness

- Understand your organization's incident response plan
- Know who to contact if ransomware is suspected
- Don't hide infections out of embarrassment
- Report suspicious emails before clicking
- Participate in security training seriously

Chapter-9

Cyber Mercenaries - The NSO Group and the Spyware Industry

"Pegasus turned our phones from tools of liberation into tools of oppression. Every journalist, every activist, every dissident now carries their own personal spy."
— **Citizen Lab researcher,** *after Pegasus Project revelations*

October 2019, WhatsApp Headquarters, Menlo Park, California

The security team at WhatsApp discovered something that shouldn't have been possible. Attackers were injecting spyware through WhatsApp calls—calls the victims never answered. The missed call would disappear from logs, leaving no trace except the malware now controlling the phone. Over 1,400 users had been targeted, including human rights defenders, journalists, and government officials. The attack used Pegasus, the crown jewel of NSO Group's surveillance arsenal.

NSO Group represents the industrialization of surveillance, transforming capabilities once exclusive to superpowers into products available to any government willing to pay. Founded in 2010 by former Israeli intelligence officers, NSO created a business model selling cyber weapons to governments worldwide. Their tagline—"Making the world a safer place"—belied the dark reality of how their tools were used.

Pegasus wasn't just spyware—it was total phone domination. Once installed, it could access everything: messages, emails, photos, passwords, location data, calendar events. It could activate cameras and microphones, turning phones into perfect surveillance devices. It operated invisibly, consuming minimal battery and bandwidth. Victims had no indication their most intimate moments were being observed.

The technical sophistication of Pegasus evolution tracked the smartphone security arms race. When Apple patched one vulnerability, NSO found another. They maintained arsenals of zero-day exploits worth millions on the black market. Their infection vectors evolved from requiring user interaction to "zero-click" attacks that needed no victim participation. They defeated end-to-end encryption by compromising endpoints themselves.

The business model was lucrative. NSO charged governments millions for Pegasus licenses, with additional fees per target. A New York Times investigation revealed Mexico paid over $60 million. Saudi Arabia reportedly paid $55 million. The company claimed to vet customers for human rights compliance, but evidence suggested profit trumped principles.

The targeting of Jamal Khashoggi's inner circle revealed Pegasus's role in human rights abuses. Before the Saudi journalist's murder in Istanbul, NSO's spyware infected phones belonging to his fiancée, friends, and colleagues. While NSO claimed no knowledge of specific targets, their tool enabled the surveillance that preceded the assassination.

The Pegasus Project, a collaboration between journalists and researchers, exposed the global scope of abuse. Analysis of leaked data suggested over 50,000 potential surveillance targets across 50 countries. The list included presidents, prime ministers, kings, and dozens of journalists. Emmanuel Macron, Imran Khan, and multiple royal family members appeared as potential targets.

The investigation revealed how governments used Pegasus to suppress dissent. Indian journalists critical of Modi's government were targeted. Hungarian reporters investigating corruption found their phones compromised. Mexican activists fighting corporate pollution were surveilled. The tool marketed for fighting terrorism and crime became a weapon against civil society.

The technical analysis of Pegasus revealed remarkable capabilities. It used exploits in iMessage, WhatsApp, FaceTime, and other common apps. It could persist through factory resets. It could extract data from cloud backups. It could even infect phones through nearby wireless connections. The sophistication suggested resources comparable to nation-state intelligence agencies.

NSO's corporate structure reflected the murky world of surveillance technology. Owned by private equity firms, registered in Luxembourg, operating from Israel, selling globally—the complex structure obscured accountability. When criticized, they claimed they were merely tool providers, not responsible for misuse. When sued, they claimed sovereign immunity as agents of foreign governments.

The legal battles surrounding NSO Group set precedents for the cyber arms trade. WhatsApp sued them in U.S. court for violating the Computer Fraud and Abuse Act. Apple sued for targeting their users. NSO argued they had sovereign immunity, essentially claiming to be part of foreign intelligence operations. Courts grappled with applying existing law to novel circumstances.

The Israeli government's role remained ambiguous. NSO's exports required government licenses, suggesting official approval. Former intelligence officers populated the company. Israeli diplomacy sometimes included offers of NSO technology. The line between private company and state asset blurred, reflecting Israel's strategy of leveraging cyber capabilities for diplomatic influence.

Competition in the spyware industry drove continuous innovation. Companies like Candiru, Quadream, and Cellebrite offered similar capabilities. The market expanded beyond phones to computers, cars, and smart home devices. Prices dropped as technology commoditized. Capabilities once costing millions became available for thousands.

The human impact of commercial spyware extended beyond direct victims. Journalists' sources feared exposure. Activists self-censored knowing they might be monitored. Lawyers couldn't guarantee client confidentiality. The mere possibility of surveillance chilled free expression and association. Trust in digital communication eroded globally.

Lessons for Security Professionals

1. Zero-Day Defense Strategies
- Assume adversaries have unknown vulnerabilities
- Implement defense-in-depth beyond patching
- Use behavioral detection not just signatures
- Isolate high-risk individuals' devices
- Regular security audits by specialized firms

2. Mobile Device Security
- Deploy mobile device management (MDM) solutions
- Restrict app installations to vetted sources
- Regular device refreshes for high-risk users
- Network monitoring for anomalous mobile traffic
- Consider using burner devices for sensitive activities

3. Executive Protection Programs
- Threat assessments for high-profile individuals
- Specialized training on mobile security
- Regular device forensics for compromise detection
- Secure communication alternatives
- Physical security integration with cyber defense

4. Supply Chain Security
- Vet all technology vendors thoroughly
- Understand vendor relationships with governments
- Monitor for indicators of compromise from vendors
- Diversify vendors to reduce single points of failure
- Include security requirements in contracts

5. Incident Response for Advanced Spyware
- Develop capabilities to detect sophisticated implants
- Plan for complete device replacement if compromised
- Legal preparation for attribution challenges
- Public relations strategies for targeted surveillance
- Coordination with platform providers (Apple, Google)

Lessons for Everyone

1. Personal Device Hardening
- Keep devices updated immediately when patches release
- Use lockdown or high-security modes when available
- Minimize app installations
- Review app permissions regularly
- Consider separate devices for sensitive activities

2. Recognizing Potential Targeting
- Unusual battery drain or data usage

- Devices running hot without cause
- Strange messages or calls
- Apps crashing repeatedly
- If you're an activist, journalist, or government official, assume targeting

3. Communication Security

- Use Signal or other encrypted messaging apps
- Verify security codes with contacts
- Be cautious of video calls from unknown numbers
- Disappearing messages for sensitive conversations
- Avoid SMS for anything confidential

4. Physical Security Integration

- Don't leave devices unattended
- Be aware of who has physical access
- Use strong PINs/passwords, not just biometrics
- Consider privacy screens
- Be cautious of device repair services

5. Risk Assessment

- Honestly evaluate if you might be a target
- Understand that association with targets increases risk
- Consider the sensitivity of your communications
- Balance security with usability
- Seek professional help if you suspect compromise

Part III: Closing Reflection

"In cyberspace, the boundaries between nations, between crime and war, between privacy and security, all become blurred. We're all inhabitants of this digital world, and we're all vulnerable."
— **Mikko Hyppönen,** *Chief Research Officer, F-Secure*

The three chapters in Part III mark a fundamental shift in our journey through the dark side of the internet—from individual actors and criminal groups to the immense power of nation-states wielding code as a weapon of war. APT28's election interference, WannaCry's global rampage, and NSO Group's surveillance capabilities reveal that cyberspace has become the fifth domain of warfare, alongside land, sea, air, and space*.

What distinguishes nation-state cyber operations is not just their sophistication but their integration with broader geopolitical objectives. Russia used hacking as a tool to attack democratic systems through the DNC. North Korea used ransomware to create a situation where every computer system across the globe became a possible hostage. The Israeli government developed spyware technology, which evolved into a complete surveillance industry that turned mobile phones into complete tracking systems. The attacks represented more than technical incidents because they functioned as state-level operations, which hackers performed using their computers.

These weren't isolated technical attacks—they were acts of statecraft executed through keyboards.

The attribution challenges these operations present have fundamentally altered international relations. When Ukrainian power grids go dark, when Iranian centrifuges spin out of control, when Saudi dissidents disappear after their phones are compromised, we may know who is responsible, but proving it to a legal standard remains nearly impossible. This ambiguity isn't a bug—it's a feature that allows nations to conduct operations just below the threshold of war while maintaining plausible deniability.

The speed and scale of these attacks defy traditional defensive thinking. WannaCry infected hundreds of thousands of computers in 150 countries within hours—faster than any biological pandemic, broader than any conventional military strike. A piece of code allegedly stolen from the NSA brought hospitals to their knees, factories to a halt, and governments to crisis mode. The weapons we build to protect ourselves became threats to everyone when they fell into the wrong hands.

The commercialization of cyberweapons through companies like the NSO Group adds another layer of complexity. Any government can now acquire capabilities once exclusive to

---

superpowers, provided it has sufficient financial resources. The same tools marketed for fighting terrorism and crime are used to suppress dissent and eliminate opponents. The private sector has become a significant player in the cyber domain, with even less oversight than traditional arms dealers.

The third section of the stories illustrates how various social boundaries have lost their previous value in society—hospitals, which once enjoyed protection during wars, now face ransomware attacks. Elections operate as the fundamental element of democratic governance, but they have evolved into digital battlegrounds where information warfare takes place.

Journalists and human rights defenders, whose work depends on source protection, found that their every communication was being monitored. The rules of engagement that took centuries to develop for conventional warfare don't exist in cyberspace.

The defensive challenges these nation-state operations present seem almost insurmountable. A hospital operating with limited IT funding must establish robust security measures, as it faces threats from advanced, military-grade ransomware attacks. How does a political party protect itself from intelligence services with unlimited resources?

A journalist who needs to protect their sources must acquire methods to conceal their information because their phone has evolved into a surveillance device.

The current security environment reveals the most significant disparity between attackers and defenders that has ever existed.

Yet Part III also reveals the limitations of cyber power. Russia executed multiple complex operations that failed to produce any specific election results. Despite WannaCry's devastating spread, it was stopped by a single security researcher registering a domain. The worldwide legal challenges to Pegasus software's advanced features have led to international condemnation since their public discovery. Cyber weapons possess strong capabilities, yet they do not possess complete power.

The responses to nation-state cyber operations have been fragmented and often ineffective. Sanctions against individuals who will never leave their home countries accomplish little. The practice of "Naming and shaming" depends on the belief that enemies will experience shame emotions. Diplomatic protests are met with denials and counter-accusations. The nuclear deterrence model, which has been effective in atomic warfare, does not apply to cyber operations because these attacks make it difficult to identify attackers and produce unpredictable responses.

International law encounters difficulties when dealing with cyber operations that do not result in physical injuries yet cause significant damage to social structures. Is election interference an act of war? Is deploying ransomware against hospitals a war crime? The sale of spyware to authoritarian regimes by businesses enables these governments to carry out human rights violations through their operations. These questions remain largely unanswered as technology outpaces the development of legal frameworks.

International law struggles to address cyber operations that do not cause physical damage because they can produce significant disruptions to social structures. Is election interference an act of war? Is deploying ransomware against hospitals a war crime? The practice of selling spyware to authoritarian regimes leads companies to participate in human rights violations. These questions remain largely unanswered as technology outpaces the development of legal frameworks.

The most disturbing revelation of Part III is how normalized these operations have become. Nation-state hacking is no longer exceptional—it is now expected. Every significant power conducts cyber operations. Many smaller nations are developing capabilities. Criminal groups operate with state protection when their activities align with national interests. We've entered an era of persistent, low-level cyber conflict that may never end.

As we prepare for Part IV's exploration of vulnerabilities that affect us all, we must recognize that nation-state capabilities eventually proliferate to criminals and terrorists. The exploits developed by intelligence agencies are often used to create ransomware. The surveillance tools sold to governments end up monitoring everyone. Extremist groups adopt the information warfare techniques pioneered by state actors. What starts as state-on-state conflict inevitably impacts civilians.

The third section of the question requires us to determine whether military operations now control the internet or if they have become more dependent on internet technology. Perhaps it's both. The digital world now functions as a combat zone that exists within the same environment as ordinary people who perform their daily routines. We're all potential combatants and collateral damage in conflicts we may not even be aware of.

The cyber operations conducted by nation-states, which are described in Part III, represent the standard operating procedure of modern times. Every government is developing offensive capabilities. Every critical system is vulnerable to attack. Every piece of information exists as a potential weapon. The current situation requires us to relinquish our belief that the internet will ever return to its previous state of being independent of international politics.

The process of defense against threats begins with identifying all possible threats. Recognizing that we're all potential targets helps us prepare. Acknowledging that cyber weapons don't respect borders or distinguish between combatants and civilian forces us to think differently about security and solidarity. The stories in Part III demonstrate how nation-state operations have become standard elements of our contemporary world, uniting digital spaces with military conflicts and protected areas with dangerous zones.

---

*   Schafer, J. (2020). International Information Power and Foreign Malign Influence in Cyberspace. International Conference on Cyber Warfare and Security, (), 423-430, XVII.

Part-IV

Vulnerabilities that hurt us all

The final part of our journey into the dark side of the internet examines vulnerabilities that affect everyone—the systemic weaknesses, human failures, and technological shortcomings that create opportunities for harm. These aren't abstract threats targeting governments or corporations; they're dangers that touch our hospitals, our children, and our essential services.

These stories force us to confront uncomfortable truths about our digital dependencies. When ransomware can stop heart surgeries, when children can become international cybercriminals, when the defenders themselves have criminal pasts, we must question the foundations of our connected world. These aren't just technical problems requiring technical solutions—they're human problems requiring human understanding.

From hospitals held hostage to teenage hackers facing life-altering consequences to the complex redemption of those who straddle both sides of the law, these final chapters examine the vulnerabilities that make us all potential victims and the human stories behind both attacks and defenses.

Chapter-10

The Hospital Hack - When Ransomware Threatens Lives

"We had to choose between paying criminals and watching patients die. That's not a choice any doctor should have to make."
— Anonymous hospital administrator, *after ransomware attack*

September 10, 2020, Düsseldorf, Germany

The ambulance carrying a critically ill woman approached Düsseldorf University Hospital at 11:30 PM, blue lights flashing through empty streets. But instead of rushing her into emergency care, the drivers received devastating news: the hospital was offline, paralyzed by ransomware. They would have to drive another 32 kilometers to Wuppertal. The woman died an hour after finally reaching medical care—potentially the first death directly attributable to ransomware.

The attack on Düsseldorf University Hospital began like countless others—through a vulnerability in commercial VPN software. The attackers, believed to be the Russian cybercriminal group DoppelPaymer, probably didn't intend to target a hospital. The ransom note was addressed to Heinrich Heine University, which shares some infrastructure with the hospital. When authorities contacted the attackers to explain they'd hit a hospital, the criminals provided a decryption key. But it was too late.

Healthcare has become ransomware's perfect victim. Hospitals can't afford downtime—lives literally depend on continuous operation. Their IT systems are complex amalgamations of decades-old equipment and cutting-edge technology. They possess valuable data—medical records, research, personal information. They're perceived as having money, whether through insurance, government funding, or desperation. Most critically, they're seen as likely to pay.

Universal Health Services (UHS), one of America's largest hospital chains, fell victim in September 2020. The attack affected over 400 facilities across the United States and United Kingdom. Hospitals reverted to paper records. Lab results were delivered by hand. Surgical schedules were maintained on whiteboards. The attack cost UHS $67 million, not counting potential lawsuits from affected patients.

The technical challenges of securing healthcare infrastructure are immense. Medical devices run operating systems decades old—Windows XP powers MRI machines costing millions. Updating requires recertification, potentially costing hundreds of thousands and taking months. Network segmentation is difficult when devices need to communicate across departments. Every security measure must be balanced against impact on patient care.

Dr. Andrea Phillips at Sky Lakes Medical Center in Oregon described the chaos during their May 2020 ransomware attack: "We couldn't access patient histories. Allergies, medications, previous conditions—all invisible. We were making life-and-death decisions with incomplete information. It was like practicing medicine in the 1980s but with 2020s patient complexity."

The ransomware groups targeting healthcare showed varying levels of conscience. Some, like Maze and DoppelPaymer, claimed they wouldn't target hospitals during COVID-19. Others saw opportunity in crisis. REvil, Ryuk, and Conti conducted devastating attacks throughout the pandemic. When criticized, they rationalized that hospitals had insurance or that governments should protect critical infrastructure better.

The human impact extended beyond immediate medical care. Cancer patients had chemotherapy delayed. Organ transplants were postponed. Psychiatric patients couldn't access medications. The stress on healthcare workers, already overwhelmed by COVID-19, was immense. Some quit, unable to handle the additional pressure of working without digital tools they'd relied on for years.

Vermont's University Medical Center attack in October 2020 demonstrated ransomware's lasting impact. Even after paying millions for recovery, effects persisted for months. Chemotherapy patients were redirected to other facilities. Payroll systems remained offline, requiring manual check writing. Medical records from the attack period remained partially inaccessible. The hospital estimated full recovery would take over a year.

The underground economy supporting healthcare attacks was sophisticated. Initial access brokers sold entry to hospital networks for tens of thousands. Ransomware-as-a-Service operators provided tools and infrastructure for a percentage of profits. Money launderers converted cryptocurrency to usable funds. Negotiators specialized in hospital victims, understanding their unique pressures and payment capabilities.

Lessons for Security Professionals

1. Healthcare-Specific Security Requirements
- Understand medical device constraints and FDA regulations
- Implement compensating controls for unpatchable systems
- Design security that doesn't impede emergency care
- Create medical device inventories with criticality ratings
- Develop vendor management programs for medical technology

2. Resilience and Continuity Planning
- Plan for complete IT failure scenarios
- Maintain paper-based backup procedures
- Regular drills including clinical staff
- Ensure backup systems are truly isolated
- Create detailed recovery prioritization plans

3. Rapid Detection and Response
- Deploy healthcare-specific threat detection
- 24/7 security operations center coverage
- Clear escalation procedures for potential ransomware

- Pre-positioned incident response resources
- Regular threat hunting in medical networks

4. Stakeholder Communication

- Develop templates for patient notifications
- Create clinical communication protocols during outages
- Prepare public relations responses
- Coordinate with health authorities
- Plan for media attention during incidents

5. Regulatory and Legal Compliance

- Understand HIPAA implications of ransomware
- Document decisions affecting patient care
- Prepare for regulatory investigations
- Consider liability for delayed/denied care
- Maintain forensic evidence while restoring operations

Lessons for Everyone

1. Patient Preparedness

- Keep personal medical records copies
- Know your medications and dosages
- Have emergency contact information written down
- Understand your rights during hospital cyber incidents
- Consider medical alert bracelets for critical conditions

2. Recognizing Affected Healthcare

- Longer wait times might indicate cyber incidents
- Paper records suggest system outages
- Redirected appointments may mean ransomware
- Ask about cyber incidents if concerned
- Have backup care plans for chronic conditions

3. Personal Health Information Protection

- Understand medical records are valuable to criminals
- Monitor for medical identity theft
- Review medical bills for fraudulent charges
- Question unexpected medical communications
- Report suspicious activity to providers

4. Supporting Healthcare Cybersecurity

- Accept security measures even if inconvenient
- Report suspicious emails or calls claiming to be from providers
- Understand that healthcare cybersecurity affects everyone
- Support funding for healthcare IT security
- Be patient during recovery from incidents

5. Community Resilience

- Know alternative healthcare facilities
- Understand emergency services might be affected

- Keep first aid supplies and medications stocked
- Build relationships with healthcare providers
- Support community emergency preparedness

Chapter-11

Greg - The Youngest Hacker's Path from Prodigy to Prison

"He wasn't a hardened criminal. He was a lonely kid who found the wrong community online. By the time we caught him, he'd stolen millions but couldn't explain why."
— **FBI Agent**, *on arresting a 15-year-old hacker*

April 15, 2019, Miami, Florida

The knock came at 6 AM, just as fifteen-year-old Graham (name changed for legal reasons) was preparing for school. FBI agents flooded into his family's suburban home, seizing computers, phones, and even his gaming consoles. His mother stood in shock as agents explained that her son, who she thought spent his time playing Fortnite, was suspected of being part of an international hacking group that had stolen millions in cryptocurrency.

Graham's journey into cybercrime began innocently enough at age eleven. Frustrated by losing in online games, he searched for cheats and exploits. He discovered forums where people shared not just game hacks but techniques for "real" hacking. His natural aptitude for understanding systems, combined with unlimited time and teenage fearlessness, created a perfect storm.

By age thirteen, Graham was participating in "SIM swapping" attacks, helping older hackers steal phone numbers to bypass two-factor authentication. He saw it as a game—voices on Discord giving him instructions, cryptocurrency appearing in wallets, respect from people who seemed important. He never met his conspirators in person, knowing them only by handles like "Wave" and "Lizard."

The technical education Graham received through criminal forums exceeded anything available in school. He learned about network protocols, cryptography, social engineering, and operational security. He could navigate Linux systems, write Python scripts, and understand blockchain technology better than many professionals. But this education came with a dark curriculum—how to steal, how to launder money, how to destroy lives.

Graham's parents, both professionals with limited technical knowledge, had no idea what their son was doing. They saw him on his computer constantly but assumed he was gaming or doing homework. When he asked for a better graphics card or faster internet, they saw it as supporting his interests. The thousands of dollars in computer equipment in his room seemed normal for a teenager interested in technology.

The escalation from minor crimes to major felonies happened gradually. First, it was just helping with SIM swaps for a few hundred dollars in Bitcoin. Then he learned to conduct the swaps himself. He moved from targeting random individuals to focusing on cryptocurrency investors. His biggest score—$800,000 from a Silicon Valley executive—made him infamous in underground circles.

The psychology of teenage hackers like Graham reveals a complex mix of motivations. The money mattered, but not as much as the respect, the thrill, and the sense of belonging. These were often lonely kids who found community online. Their crimes were abstractions—numbers on screens, not real people suffering real losses. The disconnect between actions and consequences was complete.

Law enforcement's pursuit of Graham's crew demonstrated the challenges of investigating juvenile cybercrime. The hackers were minors spread across multiple countries. They used sophisticated operational security—encrypted communications, cryptocurrency tumblers, proxy chains. But they were still teenagers, and teenagers make mistakes. Graham's came when he used stolen funds to buy designer clothes and had them shipped to his real address.

The arrest shattered Graham's family. His parents faced potential civil lawsuits from victims. His younger siblings were traumatized by the raid. His father's security clearance was revoked, costing him his job. His mother spent their savings on legal defense. The family that had been planning college visits was now visiting their son in juvenile detention.

The legal system struggled with cases like Graham's. Prosecutors wanted to charge him as an adult—his crimes were serious, causing massive financial damage. Defense attorneys argued he was a child, manipulated by older criminals, deserving rehabilitation not punishment. The judge faced an impossible decision: balance the severity of the crimes with the age of the criminal.

Graham ultimately received two years in juvenile detention followed by five years of supervised release with no internet access. For a teenager whose entire identity revolved around technology, it was a digital death sentence. He couldn't attend online school during COVID-19. He couldn't apply for jobs requiring computer use. He was effectively exiled from modern society.

The rehabilitation challenges for young hackers are immense. Traditional juvenile programs focus on drug crimes and violence, not cryptocurrency theft and computer intrusions. Counselors struggle to understand the crimes, let alone address them. The skills that made these teenagers criminals could make them valuable security professionals, but criminal records make legitimate employment nearly impossible.

Some countries have experimented with alternative approaches. The Netherlands' Hack_Right program diverts young hackers from prosecution into education programs. The UK's Cyber Choices program intervenes before crimes occur. These programs recognize that young hackers aren't traditional criminals and might benefit more from redirection than punishment.

Graham's story isn't unique. Hundreds of teenagers worldwide have been arrested for cybercrime, with thousands more engaged in illegal activities. They operate in Discord servers and Telegram channels, sharing tools and techniques. They compete for reputation and cryptocurrency. They see themselves as hackers, not criminals, until law enforcement arrives.

The role of online communities in radicalizing young hackers parallels other forms of online radicalization. Vulnerable teenagers seeking belonging find communities that reward increasingly extreme behavior. Older members groom younger ones, teaching them skills while exploiting their naivety. The progression from curious kid to criminal hacker can happen in months.

Today, Graham works at a fast-food restaurant, his computer ban finally lifted but his record making technology jobs impossible. He watches former accomplices—those who avoided arrest—working at technology companies or attending university. He sees news about cybersecurity salaries and knows he has the skills but not the opportunity. His talent, which could have protected systems, instead rots unused.

The broader implications of cases like Graham's affect cybersecurity's future. Every young hacker arrested represents both a threat neutralized and an opportunity lost. These individuals understand systems in ways that can't be taught in traditional education. Their redemption could strengthen defense. Their permanent exclusion weakens it.

Lessons for Security Professionals

1. Youth Intervention Programs
- Develop programs identifying at-risk youth before crimes occur
- Create legitimate channels for young hacking talent
- Partner with schools to provide cybersecurity education
- Establish mentorship programs for technically gifted youth
- Advocate for rehabilitation over pure punishment

2. Understanding Juvenile Threat Actors
- Recognize that young hackers may have sophisticated capabilities
- Monitor gaming and Discord communities for threats
- Understand the social dynamics driving youth cybercrime
- Develop age-appropriate incident response procedures
- Consider rehabilitation potential in incident handling

3. Parental and Educational Outreach
- Create resources helping parents recognize warning signs
- Develop a curriculum teaching ethical hacking
- Provide guidance on monitoring children's online activities
- Build relationships with schools and youth organizations
- Promote positive outlets for technical curiosity

4. Law Enforcement Coordination
- Advocate for appropriate handling of juvenile cases
- Support diversion programs over incarceration
- Help law enforcement understand youth hacker motivations
- Provide expertise in juvenile cybercrime investigations
- Push for records expungement enabling rehabilitation

5. Industry Responsibility
- Create pathways for reformed hackers to contribute
- Develop internship programs for at-risk youth
- Support organizations working with young hackers
- Consider criminal records in context
- Invest in prevention through education

Lessons for Everyone

1. Parental Awareness
- Monitor children's online activities appropriately
- Understand warning signs of hacking behavior
- Know what Discord, Telegram, and similar platforms are
- Watch for unexplained money or expensive items
- Maintain open communication about online activities

2. Recognizing At-Risk Behavior
- Excessive secrecy about online activities
- Unusual knowledge about hacking or cryptocurrency
- Bragging about illegal activities online
- Sudden acquisition of money or goods
- Association with older online "friends"

3. Positive Alternatives
- Encourage legitimate cybersecurity education
- Support participation in Capture The Flag competitions
- Find mentors in information security
- Explore bug bounty programs for appropriate ages
- Channel curiosity into constructive activities

4. Intervention Strategies
- Address concerning behavior early
- Seek professional help if needed
- Understand the legal consequences of cybercrime
- Contact law enforcement if crimes are discovered
- Focus on rehabilitation not just punishment

5. Supporting Affected Youth
- Understand that young hackers need guidance
- Advocate for second chances
- Support programs helping technical youth
- Recognize the waste of criminalizing talent
- Push for criminal justice reform in cyber cases

Chapter-12

MalwareTech - The Accidental Hero Who Stopped WannaCry

*"The same skills that make someone dangerous can make them invaluable. The question is: do we want
reformed hackers defending us, or do we exile them and hope for the best?"*
— **Security researcher**, *on the MalwareTech case*

May 12, 2017, 2:30 PM, Ilfracombe, England

Marcus Hutchins was eating lunch when a colleague messaged about a new ransomware
outbreak. The 22-year-old security researcher, known online as MalwareTech, reluctantly
returned to his computer. He'd been tracking malware for years, and ransomware was
depressingly common. But as he analyzed this new sample, he realized this was different.
WannaCry wasn't just spreading—it was exploding across the globe at unprecedented speed.

Within hours, Hutchins had inadvertently become a hero. His discovery and activation of
WannaCry's kill switch saved potentially billions in damages. But his past would soon catch up
with him. The young man celebrated for stopping a global cyber attack had a secret: years
earlier, he had created and sold malware used to steal banking credentials. His story embodies
the complex relationship between the security community and the criminal underground.

Marcus Hutchins grew up in rural Devon, England, a bright but isolated teenager who found
community online. Like many who would later enter information security, he began in the gray
areas of the internet—gaming forums where cheats were traded, IRC channels where exploits
were discussed, underground forums where the line between research and crime blurred.

By age sixteen, Hutchins had written his first malware—a simple keylogger he sold for $50. The
buyer's enthusiasm and the ease of the sale intoxicated him. Here was something he was good
at, something that earned respect and money. He improved his code, learned new techniques,
and gradually became known in underground forums as someone who could deliver quality
malware.

The malware that would later haunt him was called Kronos, a banking trojan he developed
between 2012 and 2013. Kronos was sophisticated for its time, using advanced techniques to
steal credentials from online banking sessions. Hutchins sold it for $7,000, a fortune for a
teenager living with his parents. He convinced himself he was just selling code—what buyers
did with it wasn't his responsibility.

But even as he sold malware, Hutchins was developing a parallel identity as a legitimate security
researcher. He started the MalwareTech blog, analyzing threats and sharing indicators of
compromise. His clear writing and technical insights gained followers. He tracked botnets,
sinkholed domains, and helped victims recover from infections. He was simultaneously
creating threats and defending against them.

The cognitive dissonance of his dual life took its toll. Hutchins suffered from depression and anxiety, self-medicating with drugs. He knew that discovery would destroy his legitimate career before it began. By 2014, he had stopped selling malware, focusing entirely on defensive research. But the internet never forgets, and his past remained encoded in forums and transaction records.

His work as MalwareTech gained recognition in the security community. He was invited to conferences, quoted in media, and respected by peers. His botnet tracking was particularly valuable, providing early warning of new threats. He had successfully transformed from black hat to white hat, or so he thought.

When WannaCry erupted on May 12, 2017, Hutchins was perfectly positioned to analyze it. His experience with malware internals, his understanding of criminal infrastructure, and his quick thinking led him to register the domain that served as WannaCry's kill switch. The discovery was partially luck—the domain check might have been a sandbox detection mechanism—but his decision to register it saved countless organizations.

Overnight, Hutchins became a celebrity in the security world. Media outlets worldwide celebrated the young researcher who stopped a global attack. He gave interviews, appeared on television, and was invited to speak at conferences. He attended DEF CON in Las Vegas as a hero, finally able to meet in person the researchers he'd only known online.

Then, as he prepared to fly home from Las Vegas, FBI agents arrested him at the airport. The charges related to Kronos, the banking malware he'd created years earlier. The whiplash from hero to criminal suspect was devastating. The security community rallied to his defense, raising funds for legal fees and writing character references. Many researchers had questionable pasts—the skills needed to defend often came from understanding offense.

The legal proceedings dragged on for two years. Prosecutors painted him as a malware author who profited from cybercrime. His defense emphasized his transformation and contributions to security. The stress was immense—facing potential decades in prison while maintaining his blog and research. He continued tracking threats and sharing intelligence even while under indictment.

In July 2019, Hutchins pleaded guilty to two charges related to writing and selling Kronos. The judge, recognizing his rehabilitation and contributions, sentenced him to time served with one year of supervised release. The sentence acknowledged both his crimes and his redemption, setting a precedent for how the justice system might handle reformed hackers.

The case raised fundamental questions about redemption in cybersecurity. Should past crimes permanently exclude someone from the field? How do we balance accountability with the value reformed criminals bring? The security industry needs people who understand attacker mindsets—often because they once shared them.

Hutchins returned to England and continued his security research, though his US criminal record limited opportunities. He was transparent about his past, using it to educate others about the slippery slope from curiosity to crime. His story became a cautionary tale and an inspiration—proof that people can change and contribute positively despite past mistakes.

The broader implications of the MalwareTech case affect the entire security industry. Many respected researchers have legally questionable histories. The skills that make great defenders often come from offensive experience. The industry must balance welcoming reformed individuals while maintaining ethical standards.

Today, Marcus Hutchins continues his work in cybersecurity, his past both a burden and a qualification. He represents thousands of security professionals who walked similar paths—from curious teenager to criminal to defender. His story reminds us that the line between white hat and black hat isn't always clear, and that redemption is possible for those willing to earn it.

Lessons for Security Professionals

1. Evaluating Reformed Talent

- Consider the full context of past actions
- Assess genuine rehabilitation and contribution
- Understand that offensive experience creates defensive expertise
- Balance second chances with organizational risk
- Create pathways for redemption in the industry

2. Malware Analysis Capabilities

- Develop reverse engineering skills
- Understand malware economics and infrastructure
- Build relationships with reformed researchers
- Create safe environments for malware analysis
- Share intelligence while protecting sources

3. Kill Switch and Sinkhole Operations

- Maintain infrastructure for defensive domain registration
- Understand legal implications of sinkholing
- Coordinate with law enforcement on operations
- Document defensive actions thoroughly
- Prepare for unintended consequences

4. Community and Collaboration

- Support researchers facing legal challenges
- Share threat intelligence openly
- Mentor young researchers away from crime
- Build trust across the security community
- Balance transparency with operational security

5. Ethical Guidelines

- Establish clear boundaries for research
- Document and disclose research appropriately

- Avoid crossing from research into crime
- Understand legal frameworks globally
- Promote ethical behavior through example

Lessons for Everyone

1. Understanding Security Researchers

- Recognize that defenders often have complex pasts
- Support rehabilitation and second chances
- Understand the value of reformed perspectives
- Don't demand perfection from those protecting us
- Appreciate the sacrifices researchers make

2. Recognizing Malware Indicators

- Unusual system behavior or slowness
- Unexpected pop-ups or messages
- Files with strange extensions
- Disabled security software
- Ransom messages or encrypted files

3. Supporting Defensive Research

- Participate in responsible disclosure programs
- Report suspicious activities to researchers
- Understand that security research benefits everyone
- Support funding for defensive infrastructure
- Advocate for legal protections for researchers

4. Personal Redemption

- Understand that people can change
- Support programs helping hackers reform
- Recognize the value in past mistakes
- Don't let past define future entirely
- Encourage positive contributions from anyone

5. Community Security

- Share threat information with others
- Support local cybersecurity education
- Understand that security affects everyone
- Participate in security awareness programs
- Build resilient communities through collaboration

Part IV: Closing Reflection

"I'm not the same person I was when I was 18. I've spent the last few years fighting the same malware I used to be obsessed with writing. But the industry needs to understand that you can't protect a system if you aren't allowed to understand how it breaks. We need to decide if we want our best talent in prison or on the front lines." — **Marcus Hutchins**, *Security Researcher (Source: Reflecting on his 2017 arrest after stopping WannaCry)*

The three chapters in Part IV bring our journey full circle, returning from the abstract world of nation-state operations to the immediate, personal vulnerabilities that touch our daily lives. When ransomware stops a heart surgery, when a teenager faces federal prison for keyboard crimes, when our heroes have criminal pasts—these stories force us to confront the human cost of our digital dependencies.

The attack on healthcare systems represents a moral event horizon in cybercrime. There's something fundamentally different about ransomware that prevents cancer treatment versus ransomware that steals credit cards. The criminals who encrypt hospital systems while patients suffer have crossed a line that even other criminals once respected. Yet they continue operating, often from jurisdictions that tolerate or even encourage their activities as long as they don't target local victims. The death in Düsseldorf wasn't just a tragedy—it was a preview of how cyber attacks can become literally life-threatening.

The epidemic of teenage cybercriminals challenges everything we think we know about crime and punishment. The children face this situation because their bedrooms have been transformed into crime scenes. The economy requires their skills, but we prevent them from using computers, which amounts to stopping talented musicians from playing musical instruments. The actual tragedy emerges from their criminal conduct and the fact that innocent people cannot receive forgiveness.

Modern cybersecurity presents a complex situation, as exemplified by Marcus Hutchins, because it has become impossible to distinguish between his actions when attacking and defending systems. The same person who created banking malware stopped a global ransomware pandemic. The abilities that made him dangerous also made him extremely useful.

His prosecution, while being celebrated as a hero, reveals our confused relationship with those who understand these systems well enough to both break and protect them. These three stories share a common thread: they reveal vulnerabilities that exist not in our technology but in our humanity. Hospitals are vulnerable because they must prioritize patient care over security. Teenagers become criminals because we offer no legitimate outlets for their curiosity and talent. Security professionals often have questionable pasts because you can't defend against attacks you don't understand.

Part IV also exposes the inadequacy of our current approaches to these problems. We treat cybercrime like traditional crime, but a teenager stealing millions in cryptocurrency isn't comparable to robbing a bank. We approach healthcare security with the same frameworks used for financial services, ignoring the life-and-death stakes. We demand that our defenders have spotless backgrounds while our adversaries recruit from prisons. The systemic vulnerabilities require solutions that exceed the capabilities of personal efforts. A hospital can implement perfect security practices and still fall victim to ransomware through a vulnerability in a medical device manufacturer's system. Parents can monitor their teenager's computer use and still miss the signs of criminal activity conducted through encrypted channels.

Security professionals can reform completely yet remain forever tainted by past mistakes. What's particularly troubling about Part IV's stories is their universality. Every hospital is vulnerable to ransomware. Every talented teenager could become either a criminal or a defender, depending on which community they are first found by. Every security professional faces ethical dilemmas about how their skills should be used. These aren't edge cases—they're systematic problems affecting millions.

The solutions, if they exist, require fundamental changes in how we approach cybersecurity. We need to recognize healthcare as critical infrastructure deserving special protection and resources. We need to create pathways for young technical talent that don't require breaking laws to prove capability. We need to accept that our best defenders might be reformed attackers and develop systems for redemption rather than permanent exile.

Part IV ultimately asks us to consider the kind of digital society we want to build. Do we want a world where hospitals must choose between paying criminals and watching patients die? Where do curious teenagers become federal prisoners? Where are those capable of protecting us excluded because of past mistakes? These aren't technical questions—they're moral ones that require us to balance justice with mercy, security with opportunity, punishment with rehabilitation. The stories of vulnerable hospitals, criminal teenagers, and reformed hackers aren't separate phenomena—they're interconnected symptoms of a digital ecosystem that evolved faster than our ability to govern it. The teenager arrested today could be the security researcher who saves a hospital tomorrow, but only if we create pathways for that transformation to occur. The hospital that pays ransom today funds the attack on another hospital tomorrow, perpetuating a cycle that enriches criminals while impoverishing healthcare.

As we conclude our exploration of the dark side of the internet, Part IV reminds us that we're all vulnerable, we're all potential victims, but we're also all part of the solution. The nurse who questions a suspicious email, the parent who guides their technically gifted child toward ethical hacking, the executive who hires a reformed hacker—these individual choices aggregate into societal change.

The vulnerabilities exposed in Part IV can't be patched with software updates or fixed with firewall rules. The digital era forces us to reevaluate our fundamental beliefs regarding criminal conduct, penal systems, forgiveness, and accountability practices.

They force us to acknowledge that in an interconnected world, a vulnerability anywhere is a vulnerability everywhere.

A ransomware attack resulted in severe damage to the hospital, while a teenager received federal charges, and a former hacker tried to find peace with society. This is our current reality, where global events have immediate and tangible impacts on people's lives worldwide. Part IV presents no straightforward solutions because basic solutions do not exist.

Instead, it challenges us to confront these realities and work toward a future where security doesn't come at the expense of humanity.

The dark side of the internet isn't just about technology gone wrong—it's about human systems failing to adapt to technological change. The vulnerabilities that harm us all aren't just bugs in code, but flaws in how we organize society, allocate resources, and treat one another. Fixing them requires not just better security but better humanity.

As we close this journey through cybersecurity's darkest corners, remember that every vulnerability is also an opportunity—to improve, to learn, to build better systems. The hospital attacks demonstrate that organizations need to develop resilience as a fundamental requirement for their survival. Young people need proper guidance and development opportunities, as teenage hackers illustrate the importance of this. The reformed criminals show us that redemption is possible. Learning from vulnerable stories becomes a source of strength when we demonstrate our commitment to change through determined action.

Conclusion

Lessons From Cybersecurity: Cautionary Tales

"Every hacker started as a curious kid. Every defender was once naive. Every hero has flaws. The dark side of the internet isn't inhabited by monsters—it's inhabited by humans, and that's what makes it both terrifying and hopeful." — **From the book**

As we close this journey through the Cybersecurity: Cautionary Tales, several themes emerge from these twelve stories. First, the human element remains both our greatest vulnerability and our greatest strength. Every attack, from Kevin Mitnick's social engineering to modern ransomware campaigns, exploits human trust, fear, or greed. Yet humans also defend, investigate, and rebuild after attacks.

Second, the democratization of hacking capabilities means anyone can become either an attacker or a defender. Teenagers with time and curiosity can threaten multinational corporations. Individual researchers can stop global malware outbreaks. The same tools and knowledge that enable crime also enable defense. This dual-use nature of cyber capabilities complicates every policy decision and ethical judgment.

Third, attribution and accountability remain fundamental challenges. When attackers operate from jurisdictions that won't prosecute them, when nation-states conduct deniable operations, when criminals and spies use the same techniques, traditional frameworks of law and order break down. We need new models for deterrence, prosecution, and international cooperation.

Fourth, the pace of technological change outstrips our ability to secure it. Every innovation creates new vulnerabilities. IoT devices, artificial intelligence, quantum computing—each advancement opens new attack vectors before we understand how to defend them. Security is always playing catch-up with innovation.

Fifth, the interconnectedness that defines our modern world makes everyone vulnerable to anyone. A hospital in Germany can be disrupted by criminals in Russia. An election in America can be influenced by hackers in Moscow. A phone in Israel can be compromised by spyware sold to Saudi Arabia. Geographic boundaries mean nothing in cyberspace.

Yet despite these challenges, there are reasons for hope. Security is improving, even if threats evolve faster. International cooperation, while imperfect, is strengthening. Young people who might have become criminals are finding legitimate outlets for their skills. Organizations are taking security more seriously. Individuals are becoming more aware.

The dark side of the internet will always exist as long as humans have capacities for both creation

---

and destruction. But by understanding these threats, learning from past incidents, and working together, we can build a more secure digital future. The battle between attackers and defenders will never end, but each victory, however small, makes the internet a little safer for everyone.

These stories remind us that cybersecurity isn't just about technology—it's about people. It's about the choices we make, the systems we build, and the values we embed in our digital world. Every one of us has a role to play in this ongoing struggle, whether as security professionals, technology users, or simply citizens of an interconnected world.

The dark side of the internet may be frightening, but knowledge is power. By understanding these threats, we can better defend against them. By learning from these stories, we can avoid repeating their mistakes. By working together, we can ensure that the internet's promise of connection, innovation, and progress isn't overshadowed by its potential for harm.

The internet, like any powerful tool, reflects humanity itself—capable of tremendous good and terrible evil. Our challenge is to strengthen the light while guarding against the dark. These stories from the dark side serve as both warning and guide, showing us where we've been and lighting the path toward a more secure future.

Appendix - A

Glossary of Cybersecurity Terms

Advanced Persistent Threat (APT): A prolonged and targeted cyberattack where an intruder gains access to a network and remains undetected for an extended period.

Air Gap: A security measure that isolates a computer or network and prevents it from establishing external connections.

Authentication: The process of verifying the identity of a user, device, or system.

Backdoor: A method of bypassing normal authentication to gain unauthorized access to a system.

Black Hat: A hacker who violates computer security for malicious purposes or personal gain.

Botnet: A network of infected computers controlled remotely by cybercriminals.

Brute Force Attack: A trial-and-error method of obtaining passwords or encryption keys by trying every possible combination.

Buffer Overflow: A vulnerability where a program writes more data to a buffer than it can hold, potentially allowing code execution.

Certificate Authority (CA): An entity that issues digital certificates for secure communications.

Cryptography: The practice of securing information by transforming it into an unreadable format.

DDoS (Distributed Denial of Service): An attack that overwhelms a system with traffic from multiple sources.

Deepfake: AI-generated fake videos or audio recordings that appear authentic.

Digital Signature: A mathematical scheme for verifying the authenticity of digital messages or documents.

Encryption: The process of converting information into code to prevent unauthorized access.

Exploit: A piece of software or technique that takes advantage of a vulnerabilit

Firewall: A network security system that monitors and controls incoming and outgoing network traffic.

Hashing: Converting data into a fixed-size string of characters, typically for security purposes.

Honeypot: A decoy system designed to attract and detect attackers.

IoT (Internet of Things): Physical devices connected to the internet and able to collect and share data.

Keylogger: Software that records keystrokes to capture passwords and other sensitive information.

Malware: Malicious software designed to damage or gain unauthorized access to systems.

Man-in-the-Middle (MitM): An attack where the attacker intercepts communications between two parties.

Multi-Factor Authentication (MFA): Security requiring multiple forms of verification to grant access.

Patch: A software update designed to fix vulnerabilities or bugs.

Penetration Testing: Authorized simulated cyber attack on a system to evaluate security.

Phishing: Fraudulent attempts to obtain sensitive information by disguising as trustworthy entities.

Ransomware: Malware that encrypts files and demands payment for decryption.

Root/Rooting: Gaining administrative access to a device or system.

Sandbox: An isolated testing environment for running untrusted programs.

SIM Swap: Fraudulently transferring a phone number to a new SIM card to intercept communications.

Social Engineering: Manipulating people to divulge confidential information or perform actions.

Spear Phishing: Targeted phishing attacks directed at specific individuals or organizations.

SQL Injection: Inserting malicious SQL code into application queries to manipulate databases.

SSL/TLS: Cryptographic protocols for secure communications over networks.

Tor: Software for anonymous communication through a volunteer overlay network.

Trojan Horse: Malware disguised as legitimate software.

Two-Factor Authentication (2FA): Security requiring two different authentication factors.

Virtual Private Network (VPN): Creates a secure, encrypted connection over a less secure network.

Vulnerability: A weakness in a system that can be exploited by threats.

White Hat: An ethical hacker who helps organizations find and fix security vulnerabilities.

Zero-Day: A vulnerability unknown to those who should be interested in its mitigation.

Zero Trust: Security model that requires verification from everyone trying to access resources.

Appendix - B

Resources for Security Professionals

Incident Response Frameworks

NIST Cybersecurity Framework
- Website: https://www.nist.gov/cyberframework
- Provides standards, guidelines, and best practices for managing cybersecurity risk
- Five core functions: Identify, Protect, Detect, Respond, Recover
- Industry-agnostic and scalable for organizations of all sizes

SANS Incident Response Process
- Six phases: Preparation, Identification, Containment, Eradication, Recovery, Lessons Learned
- Comprehensive training and certification programs available
- Extensive library of free resources and tools
- Website: https://www.sans.org/incident-response

ISO/IEC 27035
- International standard for incident management
- Provides structured approach to detecting, reporting, and assessing incidents
- Includes guidelines for incident response team formation
- Compatible with other ISO 27000 series standards

Professional Organizations

Information Systems Security Association (ISSA)
- Global professional organization for cybersecurity practitioners
- Local chapters provide networking and education opportunities
- Professional development and certification support
- Website: https://www.issa.org

ISACA
- Global association for IT governance professionals
- Offers certifications including CISA, CISM, and CRISC
- Extensive research and frameworks for IT governance
- Website: https://www.isaca.org

ISC2 (International Information System Security Certification Consortium)
- Offers CISSP and other prestigious certifications
- Professional development and continuing education
- Global advocacy for cybersecurity professionals
- Website: https://www.isc2.org

Technical Resources

OWASP (Open Web Application Security Project)
- Community-driven source for application security
- OWASP Top 10 vulnerabilities list updated regularly
- Free tools, documentation, and methodologies
- Website: https://owasp.org

MITRE ATT&CK Framework
- Comprehensive matrix of adversary tactics and techniques
- Based on real-world observations
- Essential for threat hunting and defense planning
- Website: https://attack.mitre.org

CVE (Common Vulnerabilities and Exposures)
- Standardized identifiers for known vulnerabilities
- Maintained by MITRE Corporation
- Critical for vulnerability management programs
- Website: https://cve.mitre.org

Training and Education

Cybrary
- Online cybersecurity training platform
- Free and paid courses covering all skill levels
- Hands-on labs and practice environments
- Career path guidance and certification prep

SANS Cyber Aces
- Free online tutorials covering security basics
- Operating systems and networking fundamentals
- Ideal for beginners and career changers
- No prerequisites required

Professor Messer
- Free CompTIA certification training videos
- Comprehensive coverage of Security+, Network+, A+
- Study groups and additional resources
- YouTube-based for easy access

Threat Intelligence Sources

US-CERT (CISA)
- United States Computer Emergency Readiness Team
- Alerts, bulletins, and vulnerability notes
- Free threat intelligence feeds
- Website: https://www.cisa.gov

AlienVault Open Threat Exchange (OTX)

- Community-powered threat intelligence platform
- Over 100,000 contributors worldwide
- Free API access to threat data
- Integration with security tools

Recorded Future

- Commercial threat intelligence platform
- Free community edition available
- Real-time threat intelligence from web sources
- Predictive analytics capabilities

Appendix - C
Digital Hygiene for Everyone

Password Security Essentials

Creating Strong Passwords
- Minimum 12 characters, ideally 16 or more
- Mix uppercase, lowercase, numbers, and symbols
- Avoid dictionary words and personal information
- Use passphrases: "Coffee@7AM!Makes$Me&Happy2024"
- Never reuse passwords across accounts

Password Manager Recommendations
- Choose reputable managers: Bitwarden, 1Password, Dashlane
- Use a strong master password you can remember
- Enable two-factor authentication on your password manager
- Regularly audit and update stored passwords
- Keep emergency access information in a secure location

Email Safety

Recognizing Phishing Attempts
- Check sender addresses carefully (look for misspellings)
- Hover over links before clicking to see actual destinations
- Be suspicious of urgent language and threats
- Verify unexpected attachments before opening
- When in doubt, contact the sender through a different channel

Email Best Practices
- Use separate emails for shopping, banking, and personal communication
- Enable two-factor authentication on all email accounts
- Regularly review account access and connected apps
- Be cautious with "unsubscribe" links in suspicious emails
- Report phishing attempts to your email provider

Social Media Security

Privacy Settings Checklist
- Review and restrict who can see your posts
- Limit who can tag you in photos and posts
- Disable location tracking and geotagging
- Be selective about third-party app permissions
- Regularly review and remove unused connected apps

Safe Sharing Guidelines
- Never post while on vacation (wait until you return)
- Avoid sharing personal information (birthdate, address, phone)
- Don't post photos of tickets, boarding passes, or documents
- Be cautious about sharing children's information and photos
- Think before posting—assume everything is permanent

Device Security

Computer Protection
- Keep operating systems and software updated
- Use antivirus software and keep it current
- Enable automatic updates when possible
- Regular backups to external drives or cloud services
- Encrypt hard drives on laptops

Smartphone Safety
- Use PIN, password, or biometric lock
- Keep phones updated with latest security patches
- Only install apps from official stores
- Review app permissions before and after installation
- Enable remote wipe capabilities

Home Network Security

Router Configuration
- Change default admin username and password
- Use WPA3 encryption (WPA2 minimum)
- Disable WPS (WiFi Protected Setup)
- Change default network name (SSID)
- Keep router firmware updated

Network Best Practices
- Create guest network for visitors
- Regularly review connected devices
- Disable remote management unless necessary
- Use VPN for sensitive activitie
- Consider network segmentation for smart home devices

Financial Protection

Banking Safety
- Never access financial accounts on public WiFi
- Set up account alerts for all transactions
- Regularly review statements for unauthorized charges
- Use dedicated computer or browser for banking
- Enable all available security features

Credit Monitoring
- Check credit reports annually (free from each bureau)
- Consider credit freezes if not actively using credit
- Set up fraud alerts after any suspicious activity
- Monitor children's credit (identity thieves target minors)
- Understand your rights under consumer protection laws

Emergency Preparedness

Before an Incident
- Document all accounts and contact information
- Set up trusted contacts for account recovery
- Know how to quickly freeze financial accounts
- Have offline backups of critical information
- Create incident response checklist

During an Incident
- Don't panic—quick, calm action minimizes damage
- Change passwords starting with financial accounts
- Contact banks and credit card companies
- Document everything with screenshots and notes
- File reports with appropriate authorities

After an Incident
- Continue monitoring accounts for months
- Consider identity theft protection services
- Update security practices based on lessons learned
- Share experience to help others avoid similar situations
- Seek support if experiencing anxiety or stress

Children's Online Safety

Age-Appropriate Guidelines
- Under 13: Direct supervision, limited access
- 13-17: Gradual independence with clear rules
- Teach critical thinking about online information
- Discuss consequences of online actions
- Model good digital behavior

Parental Controls

- Use router-level filtering for whole-home protection
- Enable parental controls on devices and apps
- Monitor screen time and app usage
- Create family technology agreements
- Keep devices out of bedrooms at night

Staying Informed

Reliable Security News Sources

- Krebs on Security (krebsonsecurity.com)
- US-CERT alerts (cisa.gov)
- Have I Been Pwned (haveibeenpwned.com)
- Electronic Frontier Foundation (eff.org)
- Your bank's security center

Red Flags to Watch For

- Unexpected password reset emails
- Unfamiliar charges on accounts
- Missing expected emails or calls
- Friends receiving strange messages from you
- Sudden loss of phone service

Remember: Perfect security doesn't exist, but good security habits significantly reduce your risk. Start with the basics and gradually improve your practices. Small steps consistently applied provide better protection than perfect security sporadically followed.

Epilogue
Cybersecurity and The Future

"The future of cybersecurity isn't about building higher walls—it's about understanding that in a connected world, we're only as secure as our weakest link. And that weakest link is usually human."
— **Bruce Schneier**, *security technologist*

…..and the strongest link is also human when we learn from our mistakes.
— **Dr. Sam Kurien**

Security Essays From My Blog

BLOG SITE

bitsbyteandbriks.com

https://bitsbytesandbricks.blogspot.com/

AI & Cybercrime

The Intersection of Artificial Intelligence and Cybercrime

The worldwide threat environment is facing a disruptive change, as cybercriminals now employ artificial intelligence (AI) systems to enhance their current attack techniques, which target autonomous threat systems.

The development of new technologies has eroded the security benefits that defenders used to maintain in detecting threats, identifying attackers, and responding to incidents.

Emerging AI-Driven Threat Vectors

AI-Enhanced Phishing and Social Engineering

AI-generated content has significantly enhanced the quality and credibility of phishing campaigns.

- **Generative language models** craft persuasive emails and text messages that mirror corporate tone, branding, and writing styles with near-perfect accuracy.
- **Deep learning algorithms** analyze public social media data and breached information to personalize attacks at scale—creating tailored spear-phishing campaigns that evade traditional awareness defenses.

These developments have effectively eliminated many of the "red flags" security teams once taught users to recognize.

Adaptive and Polymorphic Malware

AI is also enabling malware to become self-learning and adaptive.

- **Polymorphic AI-powered malware** dynamically modifies its code structure in real-time to evade signature-based antivirus tools while maintaining its core malicious behavior.
- **Machine learning exploitation engines** identify zero-day vulnerabilities at a faster rate than traditional methods, which allows attackers to convert new vulnerabilities into attack tools within short timeframes of hours instead of days or weeks. The rapid player movements create brief moments for defenders to react, as these systems operate based on predefined detection rules.

Democratization of Advanced Attack Capabilities

Perhaps the most concerning trend is the accessibility of AI-enabled offensive tools.

- **Dark web marketplaces** now offer automated vulnerability scanners, neural-network-driven password crackers, and AI-based social engineering kits.
- These tools lower the technical skill required to conduct sophisticated attacks, allowing criminal novices to execute campaigns previously associated with nation-state or advanced persistent threat (APT) groups.

As a result, the volume and velocity of AI-assisted attacks are expected to increase exponentially.

Defensive Implications: The AI Arms Race

Security operations are entering an AI-driven arms race. Defensive teams are rapidly integrating machine learning and automation to match adversarial innovation. Modern AI-powered Security Operations Centers (SOCs) employ:

- **Behavioral analytics** to detect anomalies beyond signature-based systems.
- **Automated threat hunting** to monitor for indicators of compromise continuously.
- **Adaptive defense mechanisms.** The system employs adaptive defense mechanisms that analyze each incident to develop enhanced protection strategies for future threats.
- The implementation of defensive AI systems introduces new security threats, as algorithms can develop biases and produce incorrect results, and models can become vulnerable to attacks.

Actionable Recommendations

1. Implement AI-Aware Security Training
The organization needs to update its cybersecurity awareness training to educate employees about AI-based phishing attacks, deepfake scams, and synthetic identity theft methods. The system should run simulated exercises that utilize AI to generate attack scenarios, helping users stay alert.

2. Integrate Behavioral and Anomaly Detection Systems
The system needs to use AI-based monitoring solutions that track all deviations from typical user and network behavior patterns. The solution requires human monitoring to work in conjunction with automated data analysis systems.

3. Conduct Continuous Red Team Testing
Regularly test infrastructure resilience against AI-enhanced threats. The testing process needs to perform simulations that replicate automated spear-phishing attacks, deepfake impersonation, and polymorphic malware behavior.

4. Strengthen Governance of AI and Automation Tools
Establish internal policies for the responsible use of AI in defensive systems. Require model validation, explainability, and bias auditing to prevent overreliance on automated decision-making.

5. Collaborate Across Industry and Government
Join information-sharing alliances (e.g., ISACs, CISA JCDC initiatives) to exchange intelligence on emerging AI-enabled threats and defensive best practices.

6. Prepare Incident Response Playbooks for AI Threats
The organization needs to update its response plans because AI technology creates new security threats, which include model poisoning, automated credential stuffing, and real-time deepfake impersonation.

Key Takeaway

Organizations must assume that adversaries are already using AI. The only sustainable defense is adaptation through intelligence, automation, and human-AI collaboration. Those who fail to evolve their cybersecurity posture risk being outpaced in an increasingly competitive technological landscape.

The Quantum Time Bomb

Why Your Encrypted Data Is Already At Risk

For decades, we've slept soundly knowing that our digital secrets—our bank accounts, classified communications, and blockchain data—were shielded by math so complex that solving it would require classical computers literally eons. We built a fortress on the rock-solid assumption that factoring giant prime numbers was practically impossible.

Well, folks, meet the demolition crew: the qubit.

The main issue with quantum computing extends far beyond just speed; it represents an absolute transformation in computational methods. Unlike traditional bits, quantum bits leverage superposition to exist in multiple states simultaneously. This 'quantum parallelism' enables them to explore numerous solutions simultaneously. When quantum computers leverage this immense speed advantage to factorize numbers, it creates an immediate and existential security risk.

The most famous weapon in this arsenal is Shor's Algorithm. Discovered back in 1994, it stands as the universal key that could break RSA encryption—the fundamental backbone protecting nearly all secure online communications.

The Chilling Reality: Harvest Now, Decrypt Later

The implications are staggering, and the article introduces the most frightening intelligence term today: "harvest now, decrypt later."

The current situation presents a significant data integrity issue because this threat is present in the moment. Nation-states, along with other sophisticated adversaries, continue to collect and store massive amounts of encrypted data. They know that while they can't access it today, the moment a sufficiently powerful quantum computer (one with thousands of stable, error-corrected qubits) achieves operational status, every bit of that harvested information becomes readable. This means sensitive information—including classified documents, proprietary corporate data, and financial records stretching back 20 years—will be exposed years after it was initially sent.

The Race to a Quantum-Safe Future

Fortunately, the cryptographic community is working at high speed to develop solutions known as Post-Quantum Cryptography (PQC). The entire goal here is to create cryptographic primitives based on entirely new mathematical problems that are thought to be unsolvable even for quantum computers. These new fundamental elements encompass promising fields such as lattice theory, code-based encryption, and hash-based signatures.

The recent standardization of multiple PQC algorithms by NIST marks a critical advancement, providing the crucial blueprints needed to start constructing a quantum-resistant security infrastructure.

The exact duration until the "cryptographic apocalypse" remains unclear. However, the process of transitioning our entire global network—which requires comprehensive testing and the deployment of these new algorithms across all systems—will take many years. Organizations must initiate their migration to quantum-safe security systems immediately, as delaying this process will render tomorrow's unbreakable encryption susceptible to rapid decryption. The race is on.

The Deepfake Epidemic

When Trust Becomes A Relic

Technology development follows a continuous pattern of tool creation, followed by the resulting problems that emerge from these tools. People develop tools to simplify their lives, but these tools end up creating additional complexities and security risks. The current "epidemic" attacks the core of human society because it destroys the foundation of truth. The deepfake epidemic spreads at an alarming rate because it threatens to make objective reality, which forms the basis of our institutions, become nothing more than a relic of the past.

The Bits: The Unsettling Rise of Hyper-Reality

Deepfakes represent the highest level of generative AI technology, which combines Generative Adversarial Networks (GANs) and diffusion models to achieve advanced results. The algorithms generate hyper-realistic content that makes it impossible for human eyes to detect any differences from actual reality.

The ability to produce convincing synthetic media through deepfake technology has become accessible to anyone with a standard laptop since 2025. The process of generating static deepfakes now takes only a few minutes on typical laptops, although producing real-time deepfakes with perfect audio-visual synchronization remains challenging to achieve. The technology produces exceptional results with pre-recorded content through face swapping, voice cloning, and video manipulation; however, it still exhibits noticeable flaws in lighting, micro-expressions, and temporal consistency during live deepfake operations. The widespread availability of advanced deception tools through "bits" technology has transformed the threat from theoretical to a global reality, making it accessible to anyone with malicious intentions.

The Bricks: The Corrosion of Trust

The actual impact of this technology extends beyond fake videos of public figures, as it undermines the fundamental ability to trust what we experience through our senses. The breakdown of trust in visual and auditory evidence renders it impossible to conduct democratic elections, enforce legal contracts, or verify voice calls from family members.

A finance worker at a multinational firm lost \$25 million after participating in a video conference that featured the company's CFO and other executives. The victim participated in a video conference with deepfake versions of all participants except himself, who used publicly available video footage to create their personas. The attack reached a new level of sophistication because it used multiple synthetic identities to operate in real-time.

The deepfake epidemic employs three primary methods to infiltrate the real world.

Political Destabilization: The use of manipulated videos during crucial election periods creates confusion, which weakens public trust in authentic news sources and election results.

Financial & Corporate Fraud: Synthetic voice technology enables attackers to impersonate executives, leading to employee transfers of corporate funds worth millions—the digital persona functions as an ideal weapon for crimes that involve impersonation.

Personal Injustice: The digital warfare conducted by malicious actors can result in permanent damage to the reputations of private citizens.

The Path to Resilience

Deepfake management requires three essential elements: stopping their spread, developing solutions to address them, and educating people to behave differently. The world cannot reverse the development of this technology, but we can develop protective measures for our digital systems.

The Coalition for Content Provenance and Authenticity (C2PA) offers our best hope for technical defense through its development of cryptographic standards that add tamper-evident metadata for authentic media at the time of creation. Major technology companies, including Adobe, Microsoft, and Intel, have started deploying these authentication protocols to establish a secure chain of evidence from camera sensors to consumer displays. Blockchain authentication systems provide additional security through their ability to create permanent records of authenticated content, which cannot be modified after creation.

Multiple countries have started creating new laws to control deepfake content. The EU AI Act requires all synthetic content to be labeled, while China demands explicit user permission for the production of deepfakes. The new laws create accountability pathways that hold platforms and content creators responsible for any malicious deepfake activities. The process of deepfake enforcement faces significant challenges because these fake videos spread across international borders at high speeds.

Financial institutions now use multi-factor authentication systems, which combine biometric data with challenge-response mechanisms that protect against replay attacks. News organizations have created dedicated authentication teams that treat all breaking visual content as potential deepfakes until they receive verification. The world now follows a new principle, which states that all content should be verified before acceptance, as trust has become increasingly unreliable.

The New Epistemology

The digital world enables people to create deceptive content at levels previously unimaginable. Our response needs to match this transformation by developing new methods to determine digital truth. The digital age demands that we teach media literacy as a fundamental subject starting at the elementary level and establish deepfake hygiene practices similar to those of password security, recognizing that people must actively engage with media content.

Our rapid progress in digital technology must not compromise the essential building blocks that form the foundation of our society. The way forward requires both technological solutions and a universal commitment to safeguard the shared framework of truth, which underpins all human societies.

The Inernet Of Threats

When Our Bricks Become Vulnerable

The Internet of Things (IoT) promised users a hassle-free existence through automated coffee brewing and smart home management, and automatic refrigerator inventory tracking. The digital system used billions of small sensors and continuous network connections to create a system that promised to simplify everyday tasks.

The attractive appearance of convenience technology hides an extensive security vulnerability that endangers our physical security and institutional trust. The Internet of Things (IoT) has evolved into an uncontrolled network of security weaknesses because it now contains more than 15 billion devices, which will expand to 29 billion by 2030. The Internet of Things has evolved into an unsecured network of threats that endangers our physical security and institutional trust.

The Peril of Prolific, Poorly Secured Bits

The IoT system faces two primary issues due to its economic structure and design architecture. The development process of smart devices, including smart lightbulbs and baby monitors,s focuses on fast market entry and low production costs instead of building secure systems. The manufacturing process of these devices poses multiple security risks due to the use of fixed passwords, unsecured data transmission, and a lack of secure software update capabilities. Furthermore, their supply chain is vulnerable to security weaknesses. The industrial IoT sector maintains better security measures than the consumer IoT sector, but its systems operate with outdated communication standards that were never intended for internet-based applications.

The worldwide deployment of Hikvision cameras in residential and commercial buildings continues to experience security breaches through existing backdoors and known vulnerabilities, which enable attackers to establish permanent surveillance systems. The ongoing security updates for these devices fail to prevent attackers from exploiting their backdoors and known vulnerabilities to establish ongoing surveillance systems. The Chinese manufacturer Hikvision supplies cameras that operate in critical infrastructure facilities worldwide, demonstrating how security weaknesses from a single vendor can trigger widespread system vulnerabilities.

The combination of extensive device deployment with fundamental security weaknesses has created an optimal situation for attacks. The Mirai botnet attack in 2016 showed this threat pattern but modern IoT attacks have developed into more sophisticated and enduring attacks. The current IoT security threats consist of targeted attacks that maintain their presence and operate independently.

From Data Breach to Physical Harm

The security risks associated with IoT systems differ substantially from traditional cyber threats because they pose direct threats to human life. The loss of financial data and personal identity information from large company breaches remains significant but does not typically result in fatal consequences. The direct connection between system vulnerabilities and physical damage has become a new reality in security.

An attacker successfully accessed the water treatment system in Oldsmar, Florida, during February 2021 to raise sodium hydroxide levels, which could have caused fatal poisonings for 15,000 residents. The operator's quick response saved the community from a dangerous situation that could have resulted in mass poisoning. The attack originated from remote access software that operated on an industrial control system connected to the internet through the same network as thousands of other utilities.

The healthcare industry faces an elevated risk level when it comes to security threats. The 2023 ransomware attacks on hospital IoT systems disabled essential medical equipment, including infusion pumps and ventilators, and patient monitoring systems. Medical IoT systems experience fatal consequences when they become compromised because they operate without fail-safe mechanisms. Medical facilities operated manually for several weeks due to the attacks, resulting in a noticeable deterioration of patient care.

The expansion of smart cities creates an enormous increase in security threats. The operation of traffic control systems and power grids,s and emergency services depends on sensors and actuators that maintain continuous network connections. A coordinated system attack would result in a complete shutdown of metropolitan areas, rather than just disrupting consumer services.

The Technical Reality: Not All Vulnerabilities Are Equal

The IoT threat environment shows complex characteristics. The security features of consumer devices remain minimal because they transmit data without encryption and cannot receive updates after deployment, and often contain security flaws inherent in their manufacturing components. The $20 smart plug contains fabricated chips with pre-installed backdoors that no network security measure can protect against.

Industrial and medical IoT systems operate under distinct security requirements. The transition of IT infrastructure with industrial and medical IoT systems has exposed their proprietary protocols, which were designed for air-gapped networks to internet threats. The process of updating these systems requires extensive testing, which often results in security vulnerabilities that persist for multiple years rather than short periods.

The authentication crisis exacerbates these security problems. The majority of consumer IoT devices maintain their default passwords because 85% of users have not changed them. The lack of proper certificate validation in devices makes it simple for attackers to perform man-in-the-middle attacks. These devices require hardware-based security modules to establish authentic trust relationships, as they lack this capability. Building Resilience: From Reaction to Prevention

The future demands complete transformations in IoT infrastructure deployment and management, and system design:

- **Regulatory Frameworks:** The UK Product Security and Telecommunications Infrastructure Act from 2024 establishes essential security requirements for all consumer IoT devices,s which include password protection and vulnerability disclosure, and maintenance support duration. The EU Cyber Resilience Act requires security-by-design and continuous product updates throughout all stages of product development.
- **Zero-Trust Architecture:** Organizations must treat all IoT devices as if they have already been compromised. IoT traffic runs through separate networks, which protect essential operational systems. Microsegmentation establishes separate security boundaries for different device categories. A compromised thermostat system should never enable access to medical equipment or industrial control systems.
- **Powered Defense:** Modern IoT behavior pattern monitoring systems use machine learning to identify security threats that occur when devices show abnormal activity. These systems detect threats at a faster rate than traditional signature-based methods because they handle the massive amount of continuous telemetry data from millions of devices.
- **Successful Implementations:** The implementation of IoT security measures has proven successful in specific industry sectors. Modern smart grid systems implement end-to-end encryption alongside hardware security modules and scheduled security evaluation processes. These systems have successfully defended against nation-state attacks while maintaining operational stability.

The Economics of Security

The market continues to evolve toward better security standards. Insurance providers require IoT security evaluations before issuing cyber protection policies to customers. Apple and Samsung, along with other major manufacturers, have launched security certification programs that enable businesses to differentiate their products through security features instead of basic functionality.

The core problem between security measures and affordable device prices,s and easy operation remains unresolved. Manufacturers will maintain their focus on quick market entry rather than device security, as consumers are reluctant to purchase secure products at premium prices.

The Imperative for Action

We have reached a critical juncture. The unstoppable growth of IoT devices does not mean their transformation into weapons of attack must occur. The Internet of Things can return to its original purpose through the implementation of robust security standards and defensive systems, as well as innovative approaches for managing connected devices.

The security of our digital future remains achievable because multiple successful implementations demonstrate its feasibility. The world faces a critical decision about when to take action against cyber threats, as a primary attack will eventually necessitate a response. The deployment of unsecured devices currently poses a security risk for future attacks. The physical infrastructure depends on digital ecosystem protection through immediate and continuous coordinated efforts to defend its "bits" against threats.

The Next Evolution Of Ransomware

Attacking The Integrity Of Our Bricks

Ransomware has always been a digital menace—a straightforward economic transaction of coercion. It began as a simple digital mugging, encrypting our files and demanding payment for the decryption key. It then escalated to "double extortion," where attackers not only locked the data but also stole it, threatening public release. This represented an attack on our productivity and our reputation.

The next evolution, however, targets something more fundamental: our confidence in the truth of our data.

The Emerging Threat of Data Integrity Ransomware

While not yet widespread, security researchers are observing early indicators of a concerning new tactic: Data Integrity Ransomware. Unlike traditional ransomware, which announces itself loudly through encryption, this approach operates stealthily.

In this scenario, attackers don't just encrypt; they attempt to introduce subtle modifications into critical datasets—alterations in financial ledgers, patient medical histories, or industrial control parameters. The sophistication required is significant: attackers need deep domain knowledge, bypass detection systems, and maintain persistence long enough to corrupt backups. These barriers mean we're unlikely to see widespread adoption immediately, but targeted attacks against high-value organizations are increasingly feasible.

The victim organization faces a complex decision matrix:

1. Pay the ransom: The attacker claims to provide either restoration tools or detailed change logs—though trusting criminal actors with data integrity creates its own paradox.
2. Refuse to pay: Initiate expensive forensic analysis and verification processes, potentially rebuilding systems from known-clean backups while accepting operational disruption.
3. Ignore the threat: Risk operating with potentially corrupted data, accepting liability for any downstream failures.

The economic model here is more complex than traditional ransomware. Once data integrity is questioned, trust may never fully return—making this potentially a one-shot weapon that burns the target permanently.

The Double-Edged Sword of AI Acceleration

The same AI capabilities transforming legitimate business operations will inevitably be weaponized. However, both attack and defense will be amplified:

Attack Enhancement

Malicious actors will deploy specialized AI agents for:
- Reconnaissance: LLMs analyzing public data to craft sophisticated spear-phishing campaigns
- Vulnerability Discovery: Automated scanning and exploitation of configuration weaknesses
- Persistence Maintenance: AI-driven evasion of behavioral detection systems
- Corruption Patterns: Machine learning to identify high-value data targets that maximize impact while minimizing detection

Defense Amplification

Organizations aren't defenseless. Modern security stacks include:
- File Integrity Monitoring (FIM) systems that detect unauthorized changes
- Database Activity Monitoring (DAM) tracking all modifications to critical data stores
- Cryptographic hashing and digital signatures for critical documents
- Immutable backup systems with air-gapped verification copies
- AI-enhanced SIEM platforms detecting anomalous data modification patterns

The challenge isn't that these attacks are undetectable—it's that detection and verification at scale requires significant investment in both technology and processes.

The Real Target: Institutional Trust

The true damage transcends operational disruption. When a hospital can't trust patient allergy records, when a bank questions transaction histories, when a power company doubts sensor readings—the social contract between institutions and citizens erodes.

- This erosion of trust has cascading effects:
- Regulatory scrutiny increases as authorities question data integrity
- Insurance premiums spike due to unquantifiable risk
- Transaction costs rise as every exchange requires additional verification
- Innovation slows as organizations become paralyzed by verification overhead

Consider the maritime industry's wake-up call with GPS spoofing—ships receiving falsified position data leading to groundings and collisions. Unlike the El Faro tragedy where outdated weather models proved fatal, these attacks involve actively falsified data streams. The lesson remains: our increasing dependence on data accuracy makes integrity attacks exponentially more dangerous than simple availability attacks.

Building Resilience Against Integrity Attacks

The defense isn't just technical—it's architectural and cultural:

Technical Controls
- Cryptographic provenance: Blockchain-inspired append-only logs for critical data
- Multi-party computation: Distributed verification requiring multiple compromises
- Zero Trust Data Architecture: Every data modification requires verification
- Behavioral baselines: AI systems learning normal data change patterns

Process Controls
- Change management: Every data modification tracked to an authorized source
- Segregation of duties: Critical changes require multiple approvals
- Regular integrity audits: Proactive verification rather than reactive recovery
- Incident response planning: Specific playbooks for integrity compromise scenarios

Cultural Shifts

Organizations must evolve from asking "Is our perimeter secure?" to continuously questioning "How do we verify the integrity of our operational data?" This means:
- Training staff to recognize subtle data anomalies
- Building verification steps into standard workflows
- Accepting that some efficiency must be traded for integrity assurance
- Creating clear escalation paths when data integrity is questioned

The Path Forward

Data integrity ransomware represents an evolution, not a revolution. Like previous ransomware waves, initial attacks will target unprepared organizations before defenses catch up. The organizations that survive will be those that:
- Invest proactively in integrity verification infrastructure
- Maintain offline verification capabilities for critical data
- Build response plans specifically for integrity incidents
- Create data governance frameworks that prioritize integrity alongside availability
- Foster security cultures where questioning data integrity is encouraged, not dismissed

The bricks of our digital reality—our core data—must be protected not just from theft or encryption, but from the more insidious threat of corruption. As we build increasingly automated and interconnected systems, the integrity of our data becomes the integrity of our decisions. In this new threat landscape, paranoia about data integrity isn't pathological—it's prudent.

The question for every security leader is no longer "When will we be hit by ransomware?" but rather "How will we know if our data can still be trusted when we are?"

Your Identity Was Already Stolen

I've been thinking a lot about exhaustion lately. Not the kind that comes from long hours or complex projects—the kind that attackers are deliberately weaponizing against us. And it's working.

The Attack That Exploits Human Nature

MFA fatigue attacks represent something genuinely clever in cybersecurity: an adversary who understands that the weakest point in any security system isn't the cryptography or the firewall. It's the person at 11 PM who just wants the notifications to stop.

Here's how it works. An attacker steals your credentials—probably through phishing, possibly from one of the countless breaches that have exposed nearly every American's personal information over the past two years. They attempt to log in. Your phone buzzes with an MFA push notification. You decline it. Another notification. You decline again. Then another. And another. Midnight comes. You're exhausted. The notifications keep coming.

At some point, a significant percentage of people just approve the request to make it stop.

The psychology is devastating in its simplicity. We've trained users to respond to prompts. We've built muscle memory around tapping "Approve." And attackers have figured out how to weaponize that conditioning.

The Transformations Worth Noting

This brings me to a sidebar that matters more than it might seem.

Kevin Mitnick passed away in July 2023 from cancer. For those who don't know the name, Mitnick was once the most wanted computer criminal in the United States—a social engineering pioneer who served five years in federal prison. What's worth remembering isn't just his criminal past but his transformation into one of the most respected white-hat security consultants in the industry.

One of my book's reviewers and a close friend, Andrew Starvitz, met Kevin Mitnick. He had a business card that was a metal lockpick set—perfectly on-brand for someone who spent his career demonstrating that most security is theater if you understand human nature.

Andrew Starvitz also met Frank Abagnale Jr. at a Novell NetWare event, which dates us considerably. Abagnale's story—immortalized in "Catch Me If You Can"—follows a similar arc: extraordinary criminal capability redirected toward protecting the systems he once exploited.

These transformations matter because they remind us that understanding the attacker's mindset isn't just academically interesting. It's essential. The best defenders often think like the people trying to get in.

Speaking of transformations and justice, Ross Ulbricht—founder of the Silk Road marketplace—received a full and unconditional pardon from President Trump in January 2025 after serving over a decade of his double life sentence plus forty years. Whatever your opinion on the original case, Ulbricht's release reflects the ongoing national conversation about proportional sentencing in technology crimes.

The Breach That Affected Everyone

But here's the development that should keep you up at night: the National Public Data breach of 2024.

NPD was a data broker—a company that collects, aggregates, and sells your personal information without your permission and largely without your knowledge. A cybercriminal known as "USDoD" compromised their databases starting in late 2023, ultimately exposing approximately 2.9 billion records affecting over 272 million people. Names. Addresses. Social Security numbers. Phone numbers. Emails.

The company didn't publicly confirm the breach until August 2024, months after the data had already been circulating on the dark web.

The owner of Jerico Pictures, Inc.—which does business as National Public Data—is Salvatore Verini, Jr., a former Florida law enforcement officer. This man was entrusted with hundreds of millions of Americans' most sensitive personal information, stored it on inadequately secured systems, and faced no meaningful criminal consequences when it was stolen and published.

Let me be direct: in the past two years, almost every American has had their personally identifiable information compromised through no fault of their own. Our government has been painfully slow to protect consumers. Data brokers continue to operate with minimal regulation, peddling our personal information with no consequences and no meaningful way for consumers to opt out.

You can check whether your data was exposed at npd.pentester.com. Spoiler: it probably was.

The Path Forward: Passkeys

So what about this? For individuals, passkeys represent the most significant improvement in authentication security we've seen in decades.

Passkeys replace passwords entirely with a public-private key pair tied to your device. When you authenticate, your phone or computer uses biometric verification—your fingerprint, your face, your PIN—to unlock a private key that proves your identity. The private key never leaves yourt device. There's no password to steal. There's no credential to stuff into automated attack tools.

The security properties are genuinely impressive: passkeys are phishing-resistant because they're cryptographically bound to specific websites. They eliminate the entire category of credential reuse attacks. And they're more convenient than passwords because you're using authentication methods you already use to unlock your phone.

Major platforms support passkeys now. Google, Apple, Microsoft, and most significant services have implemented them. The adoption curve is still early, but this is the direction authentication is heading.

The Bottom Line

MFA fatigue attacks work because we've created a security ecosystem that depends on human vigilance without accounting for human limitations. Data brokers have built an industry on collecting and selling information that enables such attacks, while regulatory frameworks lag years behind the threat landscape. And the breaches keep coming.

The defensive posture I'd recommend: enable passkeys everywhere you can. Use number-matching MFA where passkeys aren't available. Never approve an authentication request you didn't initiate—if you're being bombarded with notifications, that's the attacker at work, not a technical glitch. And assume your personal information has already been compromised, because statistically, it almost certainly has.

We're operating in an environment where identity theft isn't a risk to manage; it's a reality to navigate. The question isn't whether your data is out there. It's how effectively you can limit the damage.

What's your experience with passkey adoption?

Have you encountered MFA fatigue attacks in your organization?

Are the tools we're deploying keeping pace with the threats we're facing?

Love to hear your comments,

drop me a note

dr.samkm@protonmail.ch

Endorsements

"This book brings the world of cybersecurity to life. The cases are well told, technically sound, and grounded in the human side of the work. Anyone who wants to understand how attacks occur and why they persist will find value in these pages."

— Christopher Mosby

CEO, Movaci

CISSP, CCSP, CISA, CEH, CHFI, CASP+, PCIP

"Cybersecurity: Cautionary Tales is an essential guide for understanding how breaches really happen—and how to prevent them. Sam combines historically accurate case studies with decades of real-world expertise, translating complex incidents into practical, actionable guidance. The book's clear structure and mitigation checklists make it invaluable for CISOs, executives, and security professionals alike. Accessible, insightful, and deeply relevant, this should be required reading for anyone serious about cybersecurity."

— Andrew Starvitz, CISSP, CISM, ITIL

Founder & President, Touchpoint Cyber, LLC | Former CISO

"Cybersecurity: Cautionary Tales is a timely and transformative work that bridges the gap between technical expertise and everyday digital responsibility. By delivering clear lessons for both security professionals and the general public, the book makes complex cybersecurity threats accessible and deeply relevant. Its exploration of ransomware, digital hygiene, AI-driven cybercrime, and identity risk powerfully highlights the need for proactive and ethical security practices. This is an essential read for anyone committed to building a safer, more responsible digital world."

— Felix Rajan Meri Cruze, CTO

Boscosoft Technologies

Cybersecurity: Cautionary Tales powerfully reframes cybersecurity as a human and governance challenge, not merely a technical one. Through globally grounded case studies, Dr. Sam Kurien shows how trust, leadership, ethics, and preparedness shape cyber risk across societies. The book connects real-world incidents to public policy, organizational accountability, and social impact with clarity and depth. This is an essential guide for leaders, policymakers, and citizens working to build resilient and trustworthy digital ecosystems."

— Dr. Thaddeus Singarayan

Managing Director & CEO, Bosco Soft Technologies Pvt. Ltd.